Ocean–Atmosphere–Land Interactions and Their Roles in Climate Change

Ocean–Atmosphere–Land Interactions and Their Roles in Climate Change

Guest Editors

Biyun Guo
Chunyan Li

MDPI

Basel • Beijing • Wuhan • Barcelona • Belgrade • Novi Sad • Cluj • Manchester

Guest Editors

Biyun Guo
Marine Science and
Technology College
Zhejiang Ocean University
Zhoushan
China

Chunyan Li
Department of Oceanography
and Coastal Sciences
Louisiana State University
Baton Rouge, LA
USA

Editorial Office
MDPI AG
Grosspeteranlage 5
4052 Basel, Switzerland

This is a reprint of the Special Issue, published open access by the journal *Atmosphere* (ISSN 2073-4433), freely accessible at: https://www.mdpi.com/journal/atmosphere/special_issues/32JX05YZ4L.

For citation purposes, cite each article independently as indicated on the article page online and as indicated below:

Lastname, A.A.; Lastname, B.B. Article Title. *Journal Name* **Year**, *Volume Number*, Page Range.

ISBN 978-3-7258-7326-5 (Hbk)
ISBN 978-3-7258-7327-2 (PDF)
https://doi.org/10.3390/books978-3-7258-7327-2

Contents

atmosphere

MDPI

Editorial

Ocean–Atmosphere–Land Interactions and Their Roles in Climate Change

Biyun Guo [1,2,3,*] and Chunyan Li [4]

1. Marine Science and Technology College, Zhejiang Ocean University, Zhoushan 316022, China
2. Key Laboratory of Water Ecology Remediation and Protection at Headwater Regions of Big Rivers, Ministry of Water Resources, Xining 810016, China
3. School of Civil Engineering and Water Resources, Qinghai University, Xining 810016, China
4. Department of Oceanography and Coastal Sciences, Louisiana State University, Baton Rouge, LA 70803, USA; cli@lsu.edu

* Correspondence: biyunguo@163.com

The Earth functions as an integrated system in which the ocean, land, and atmosphere are connected through complex exchanges of matter and energy. Their interactions drive the evolution of global and regional climate by regulating both the water cycle and the planetary energy balance. A deep understanding of the physical, chemical, and biological processes that shape these ocean–land–atmosphere interactions is essential for accurately predicting climate change, assessing its environmental impacts, and formulating effective response strategies.

Under ongoing global climate variations, these interactions are undergoing significant transformations, manifested through a series of extreme weather and climate events. To reveal the evolving dynamics of this complex system and to showcase recent scientific advances, we organized this Special Issue. It brings together original research spanning climatology, oceanography, land surface hydrology, atmospheric science, and environmental science, collectively examining the physical, chemical, and biological processes that govern ocean–land–atmosphere interactions and their climatic effects. This Special Issue comprises eight original articles covering the following topics:

The paper "Atmospheric CO_2 Isotopic Variations, with Estimation of Ocean and Plant Source Contributions" analyzes atmospheric CO_2 concentrations and accompanying $\delta^{13}C$ measurements at ten latitudes from 90° S to 82° N [1]. Atmospheric CO_2 is separated into deep ocean and plant (including fossil fuel) sources. While the isotopic signatures indicate nearly equal contributions on average over 1978–2015, atmospheric CO_2 concentrations continue to rise. Seasonal comparisons show no change in the ocean component—peaking at the equator—and no significant spatial variation. In contrast, the plant component correlates with the timing of El Niño events and exhibits a shift associated with the 1995 Atlantic Multidecadal Oscillation (AMO) phase transition. These variations are consistent with changes in fossil fuel emissions, oceanic primary productivity, and forest and peat fires. The authors propose that changes in ocean-source CO_2 may result from upwelling of CO_2-rich deep waters transported by the global thermohaline circulation.

Water temperature is a fundamental factor regulating the ecological environments of rivers, lakes, and oceans. Under the increasing frequency of global extreme weather events, anomalous thermal patterns in inland waters are becoming more prominent [2–4]. Using the radiative transfer equation, Kong et al. [5] analyzed radiation, precipitation, wind speed, and air temperature data from cities around Poyang Lake. They found a strong correlation between inland water surface temperature and regional air temperature. Slight interannual

 https://doi.org/10.3390/atmos16121382

differences in water temperature were observed between 2020 and 2021, showing a cooling trend associated with the concurrent La Niña event. From August to October 2020 and in August 2021, northeastern parts of the lake were warmer than southwestern regions, with coastal zones generally exhibiting elevated temperatures. Persistent thermal discharge from factories along the northern shore was identified as a major contributor to both elevated temperatures and spatial heterogeneity.

Kang et al. [6] investigated atmospheric moisture transport pathways and exchange between the source regions of the Yangtze and Yellow Rivers. They found that the Bay of Bengal pathway and the Northwest China pathway play dominant roles in both PM (daily precipitation $\geq$ 10 mm) and PL (daily precipitation < 10 mm) events in the source region of the Yangtze River (SYZR), contributing 41.87% and 39.12% of PM events and 41.33% and 33.16% of PL events, respectively. In the Yellow River source region (YSR), the Bay of Bengal, Northwest China, and Southeast China pathways all contribute significantly. The net moisture flux from the SYZR to the YSR is approximately 60 times larger than the reverse, accounting for about 6.9% of total YSR precipitation.

Net Primary Productivity (NPP)—the carbon fixed by photosynthesis minus plant respiration—is a critical measure of ecosystem health and a key parameter in global carbon-cycle studies [7,8]. Li et al. [9] found that precipitation is the dominant driver of NPP variability, with temperature having no significant effect and radiation showing a negative correlation. The contributions of precipitation, temperature, and radiation to NPP change are approximately 91%, 5.3%, and 3.8%, respectively. In high-coverage meadows, NPP is mainly influenced by precipitation, followed by temperature and radiation. In low-coverage meadows, precipitation remains the primary driver, while temperature increases exert an inhibitory effect.

Sea surface temperature (SST) is a fundamental parameter governing ocean dynamics and plays a decisive role in regulating the global climate and maintaining marine ecosystem stability [10–13]. Accurate SST prediction provides essential support for marine meteorology, climate studies, fisheries, and environmental protection [14–16]. Ji et al. [17] developed an SST prediction model for the East China Sea using a random forest algorithm combined with the SA-ConvLSTM model. Through parameter importance analysis and correlation evaluation, they identified 2 m air temperature (T2m) and longwave radiation (LWR) as the key auxiliary variables. Four input datasets (SST-only, SST–T2m, SST–LWR, SST–T2m–LWR) were tested, all capturing SST variations effectively. The model showed the highest RMSE and MAE in summer and the lowest in winter.

Typhoons (hurricanes) strongly affect air–sea interactions by altering SST and modulating chlorophyll-a (Chl-a) concentrations. Responses vary across different events and spatial scales [18]. Shi et al. [19] examined SST and Chl-a responses along the track of Typhoon Lekima (2019). SST cooling decreased radially outward from the storm center and was closely linked to the typhoon's position, track, and influence area. After landfall (15–31 August), both Chl-a and SST showed a synchronous increasing trend. Their analysis suggested that stronger typhoons with longer residence times contributed more significantly to increases in Chl-a concentration.

El Niño remains one of the most influential interannual climate phenomena [20]. While many studies have documented a rising frequency of Central Pacific (CP) El Niño events, their governing mechanisms remain uncertain. Jia et al. [21] analyzed CP El Niño events from 1960 to 2022 and found a marked increase in their occurrence since 2000: eight events during 1960–1999 and six events during 2000–2022. The recurrence interval shortened from 4–5 years to 2–3 years. The increased frequency is closely associated with persistent warm SST anomalies in the central equatorial Pacific (5° S–5° N, 170° W–122° W) during the developing phase (June–October).

Anthropogenic activities and climate change have jointly altered surface water resources, rainfall patterns, and temperature trends, underscoring the importance of improved flood forecasting [22,23]. Mohammadi et al. [24] analyzed 30 years of hydrological data (1986–2018) for the Zarrinehrud River, including monthly peak discharge, precipitation, and temperature. They found that dam operations significantly modify the natural flow regime, altering precipitation–temperature–runoff relationships. Peak precipitation has shifted toward spring (April–May upstream, May–June downstream), and precipitation and discharge peaks occur earlier upstream than downstream. Stronger precipitation–discharge correlations downstream reflect cumulative inputs from tributaries and seasonal rivers.

Together, the studies in this Special Issue provide a comprehensive exploration of ocean–atmosphere–land interactions across multiple disciplines. Collectively, they offer an improved scientific foundation for understanding climate change, assessing the impacts of human activities, and informing future climate projections, thereby supporting global climate-risk assessments and the development of effective response strategies.

Author Contributions: Conceptualization, B.G.; investigation, B.G.; writing—review and editing, B.G. and C.L.; supervision, B.G. All authors have read and agreed to the published version of the manuscript.

Funding: This work was mainly supported by the CRSRl Open Research Program (Program SN: CKWV202592); the Key R&D projects in Zhejiang Province (No. 2023C03120); the Open Research Fund Program of the Key Laboratory for Water Ecology Management and Protection in River Source Areas, Ministry of Water Resources (No. 2024slbjh02); the National Natural Science Foundation of China (No. U2243226); the Consultation and Evaluation Program of the Department of Chinese Academy of Science (No. 2020-ZW11-A-023); and the Science and Technology Program of Department of Natural Resources of Zhejiang Provincial, China (No. 2024ZJDZ039).

Institutional Review Board Statement: Not applicable.

Informed Consent Statement: Not applicable.

Conflicts of Interest: The authors declare no conflicts of interest.

References

1. Quirk, T.; Asten, M. Atmospheric CO_2 isotopic variations, with estimation of ocean and plant source contributions. *Atmosphere* **2023**, *14*, 1623. [CrossRef]
2. Sun, J. Temporal and Spatial Variation of Lake Water Temperature in the Lower Reaches of the Yangtze River and Its Response to Climate Warming. Master's Thesis, Tianjin University, Tianjin, China, 2021. (In Chinese)
3. Li, Q. Variation characteristics of surface water temperature in Qinghai Lake and its response to climate change. *Yellow River* **2018**, *40*, 25–29. (In Chinese)
4. Wang, X.; Shi, K.; Zhang, Y.; Qin, B. Response of lake surface water temperature to climate change: Past, present and future. *Chin. Sci. Bull.* **2023**, *68*, 3002–3004. (In Chinese) [CrossRef]
5. Kong, X.; Li, Y.; Wang, L.; Liu, H. Lake surface temperature retrieval study based on Landsat 8 satellite imagery—A case study of Poyang Lake. *Atmosphere* **2024**, *15*, 428. [CrossRef]
6. Kang, B.; Wei, J.; Ayantobo, O.O.; Yang, H. Determination of transport pathways and mutual exchanges of atmospheric moisture between source regions of Yangtze and Yellow River Basins. *Atmosphere* **2024**, *15*, 524. [CrossRef]
7. Liu, C.; Dong, X.; Liu, Y. Changes of NPP and their relationship to climate factors based on the transformation of different scales in Gansu, China. *Catena* **2015**, *125*, 190–199. [CrossRef]
8. Shi, Z.; Wang, Y.; Zhao, Q.; Zhang, L.; Zhu, C. The spatiotemporal changes of NPP and its driving mechanisms in China from 2001 to 2020. *Ecol. Environ. Sci.* **2022**, *31*, 2111–2123.
9. Li, M.; Li, Q.; Xue, M. Spatio-temporal changes of vegetation net primary productivity and its driving factors on the Tibetan Plateau from 1979 to 2018. *Atmosphere* **2024**, *15*, 579. [CrossRef]
10. Herbert, T.D.; Peterson, L.C.; Lawrence, K.T.; Liu, Z. Tropical ocean temperatures over the past 3.5 million years. *Science* **2010**, *328*, 1530–1534. [CrossRef] [PubMed]

11. Bouali, M.; Sato, O.T.; Polito, P.S. Temporal trends in sea surface temperature gradients in the South Atlantic Ocean. *Remote Sens. Environ.* **2017**, *194*, 100–114. [CrossRef]
12. Yao, S.L.; Luo, J.-J.; Huang, G.; Wang, P. Distinct global warming rates tied to multiple ocean surface temperature changes. *Nat. Clim. Change* **2017**, *7*, 486–491. [CrossRef]
13. Lin, X.; Shi, J.; Jiang, C.; Liu, Z. The influence of ocean waves on sea surface current field and sea surface temperature under the typhoon background. *Mar. Sci. Bull.* **2018**, *37*, 396–403.
14. Nagasoe, S.; Kim, D.-I.; Shimasaki, Y.; Oshima, Y.; Yamaguchi, M.; Honjo, T. Effects of temperature, salinity and irradiance on the growth of the red tide dinoflagellate *Gyrodinium instriatum* Freudenthal et Lee. *Harmful Algae* **2006**, *5*, 20–25. [CrossRef]
15. Li, S.; Goddard, L.; DeWitt, D.G. Predictive Skill of AGCM Seasonal Climate Forecasts Subject to Different SST Prediction Methodologies. *J. Clim.* **2008**, *21*, 2169–2186. [CrossRef]
16. Solanki, H.U.; Bhatpuria, D.; Chauhan, P. Signature analysis of satellite derived SSHa, SST and chlorophyll concentration and their linkage with marine fishery resources. *J. Mar. Syst.* **2015**, *150*, 12–21. [CrossRef]
17. Ji, Q.; Jia, X.; Jiang, L.; Xie, M.; Meng, Z.; Wang, Y.; Lin, X. Contribution of atmospheric factors in predicting sea surface temperature in the East China Sea using the Random Forest and SA-ConvLSTM Model. *Atmosphere* **2024**, *15*, 670. [CrossRef]
18. Chen, D.; He, L.; Liu, F.; Yin, K. Effects of typhoon events on chlorophyll and carbon fixation in different regions of the East China Sea. *Estuar. Coast. Shelf Sci.* **2017**, *194*, 229–239. [CrossRef]
19. Shi, Y.; Guo, B.; Niu, Y.; Mantravadi, V.S.; Wang, J.; Ji, Z.; Che, Y.; Ye, M. Response of sea surface temperature and chlorophyll-a to Typhoon Lekima (2019). *Atmosphere* **2024**, *15*, 919. [CrossRef]
20. Newman, M.; Shin, S.I.; Alexander, M.A. Natural variation in ENSO flavors. *Geophys. Res. Lett.* **2011**, *38*, L14705. [CrossRef]
21. Jia, L.; Guo, Y. Increased frequency of Central Pacific El Niño events since 2000 caused by frequent anomalous warm zonal advection. *Atmosphere* **2025**, *16*, 654. [CrossRef]
22. Kordi, F.; Hamzeh, S.; Atarchi, S.; Alavipanah, S.K. Agricultural product classification for optimal water resource management using the data time series of Landsat8. *Iran. Ecohydrol.* **2018**, *5*, 1267–1283.
23. Kazemi Garajeh, M.; Wenget, O.; Haghi, V.; Li, Z.; Garajeh, A.; Salmani, B. Learning-based methods for detection and monitoring of shallow food-affected areas: Impact of shallow food spreading on vegetation density. *Can. J. Remote Sens.* **2022**, *48*, 481–503. [CrossRef]
24. Mohammadi, F.; Pu, J.H.; Guo, Y.; Hanmaiahgari, P.R.; Mohammadi, O.; Mohammadi, M.; Al-Qadami, E.; Mohammad Razi, M.A. Hydro-climatic variability and peak discharge response in Zarrinehrud River Basin, Iran, between 1986 and 2018. *Atmosphere* **2025**, *16*, 681. [CrossRef]

atmosphere

Article

Atmospheric CO_2 Isotopic Variations, with Estimation of Ocean and Plant Source Contributions

Tom Quirk [1,*] and Michael Asten [2,†]

[1] Australian Environment Foundation, P.O. Box 185, Melbourne, VIC 3126, Australia
[2] Earth Insight, 8 Rae St, Hawthorn, Melbourne, VIC 3122, Australia; michael.asten.monash@gmail.com
* Correspondence: twquirk39@gmail.com
† Retired. Formerly School of Earth Atmosphere and Environment, Monash University, Melbourne, VIC 3800, Australia.

Abstract: This analysis uses both atmospheric CO_2 concentrations and the accompanying $\delta^{13}C$ isotopic measurements of CO_2 over 40 years from 1978 to 2015 observed at ten different latitudes from 90° S to 82 °N. Atmospheric CO_2 is separated into two components of CO_2 attributable to deep ocean and to plant (including fossil fuel) sources. The isotopic values assigned to the two components are $\delta^{13}C = 0‰$ and $-26‰$, respectively. The latitude variations in residual source component CO_2 show the ocean source component peaking at the equator. This contrasts with the residual plant source component that peaks in the Arctic Circle region. Seasonal comparisons show no change in the ocean component peaking at the equator and no significant changes in its variation with latitude, while the plant component shows seasonal changes of the order of 15 ppm at high latitudes. The ocean component shows clear anomalous behavior in the three years following the 1989 Pacific Ocean regime shift (a shift independently identified from the changed biological time series). By contrast, the residual plant component shows a correlation in the timing of maxima in its annual variations with the timing of El Nino events over the time span of 1985–2015. It also shows a discontinuity in annual variation coinciding with the 1995 AMO phase change. We conclude that the ocean and plant components of atmospheric CO_2 relate to independent sources of atmospheric CO_2 and have approximately equal magnitudes. The observations are consistent with a hypothesis that variations in the ocean components have an origin from upwelling water from deep ocean currents, and variations in plant components are dominated by a combination of fossil fuel CO_2, phytoplankton productivity, and forest and peat fires, which primarily occur in the northern hemisphere.

Keywords: ^{13}C isotopes in CO_2; plant component; ocean component; independent sources of CO_2; Pacific Ocean regime shift; El Nino events

1. Introduction

It has been previously shown that the apparent smooth and continuous rise in atmospheric CO_2 concentrations can be broken into a series of trend changes associated with ocean decadal phase changes [1]. We extend that association through the separation of CO_2 concentrations on the basis of carbon isotopes in order to distinguish between variations in CO_2 concentrations associated with Pacific Ocean regime changes and variations associated with sea-surface temperature variations as represented in the Nino 3.4 index [2].

This paper does not study the bidirectional carbon cycle, which involves the bidirectional flux of CO_2 from land to atmosphere, and ocean to atmosphere (see, for example, [3,4]). Rather, we use the observed isotopic composition of CO_2 observed at ten globally distributed stations in order to obtain estimates of annual and decadal variations in atmospheric CO_2 attributable to plant origins (both natural and anthropogenic processes) and to CO_2 transfer into the atmosphere from upwelling deep ocean waters. This global perspective is thus different from, for example, the work of [5] who used carbon isotope studies on plant matter from a single location (Baragui River, Parana, Brazil) in order to deduce local climatic and vegetation composition changes over the past 400 years.

The measure of the ratio of ^{12}C to ^{13}C found in a sample of CO_2 is expressed as $\delta^{13}C$, where the $^{13}C/^{12}C$ abundance ratio is of the order of 0.01. Variations in $\delta^{13}C$ are attributable to variations in the molecular weight of CO_2, which affect the rate of chemical reactions and photosynthesis.

There are now about 60 years of atmospheric CO_2 concentration measurements available from the Scripps Institution of Oceanography (SIO) [6] and 40 years of $\delta^{13}C$ isotope measurements of CO_2 commencing in 1978. The SIO stations and years with both CO_2 and $\delta^{13}C$ measurements are listed in Table 1. The additional column shows the percentage of months with no data from 1986 to 2015. In the following analysis, we select the time period of 1986–2015 as having the most complete records. In particular, we avoid a 32-month period from 2016 to 2018 on Christmas Island, which has no data.

Table 1. SIO stations and years for CO_2 and $\delta^{13}C$ measurements, from [6].

	Latitude	Longitude	Elevation (meters)	Years with CO_2 and $d^{13}C$	% Months with no Data	
					All	1986–2015
Alert	82° N	63° W	210	1986–2018	4.55%	4.72%
Point Barrow	71° N	157° W	11	1983–2018	4.17%	4.72%
La Jolla	33° N	117° W	10	1979–2018	14.38%	10.28%
Kumukahi	19° N	155° W	3	1981–2018	1.10%	0.28%
Mauna Loa	19° N	156° W	3397	1981–2018	2.19%	0.00%
Christmas Island *	2° N	157° W	2	1978–2018	32.11%	26.67%
American Samoa	14° S	171° W	30	1985–2018	3.92%	3.33%
Kermadec	29° S	178° W	2	1985–2018	53.43%	51.94%
Baring Head	41° S	175° E	85	1986–2018	40.91%	39.17%
South Pole	90° S		2810	1978–2018	7.52%	3.06%

* Note that Christmas Island is part of the Republic of Kiribati in the Pacific Ocean.

This analysis uses the CO_2 concentration and isotope measurements after 1978 to divide the atmospheric CO_2 concentration into an "ocean" and a "plant" component by assuming values for the isotopic composition for the two components. The purpose of this study is to demonstrate a means of computing the fraction of gaseous CO_2 in atmospheric samples that can be attributed to a plant origin.

The isotopic approach used in this paper allows a phenomenological study of atmospheric CO_2 origins from upwelling deep ocean sources, and from plant sources including both anthropogenic and natural sources that release CO_2 into the atmosphere. This approach does not require assumptions or estimations of the dissolved inorganic carbon content of the ocean surface waters, nor does it quantify the uptake by the oceans of atmospheric CO_2. It does, however, provide quantitative data on seasonal, annual and latitudinal variations in fractions of atmospheric CO_2 attributable to ocean and plant sources.

2. Method, Model and Data

2.1. Deep Ocean and Plant Contributions to CO_2

The basis of this analysis is that contributions to CO_2 derived from deep ocean and from plant sources take discrete and different values. For SIO $\delta^{13}C$ data, the standard ratio $R(^{13}C/^{12}C)_{std}$ is from the Vienna Peedee Belemnite (VPDB) isotope [7] where $R_{std} = 0.011180$.

Deviations in the isotope ratio as measured in air samples are expressed as

$$\delta^{13}C_{meas} = ((R_{meas}/R_{std}) - 1) \times 1000.$$

For this analysis, we use two reference points as follows. For deep oceans:

$$^{13}C_{ocean} = 0‰$$

as given by [8–10], and for plants, we use

$$\delta^{13}C_{plant} = -26‰.$$

The plant value is representative of C3 plants [11]; phytoplankton for north and south, high latitudes [12]; and fossil fuels [13,14].

The separation of deep ocean and plant contributions can be quantified using the two relations defined in Equations (1) and (2):

$$A_{meas} = A_{ocean} + A_{plant} \tag{1}$$

$$\delta^{13}C_{meas} = (\delta\,^{13}C_{ocean} \times A_{ocean} + \delta^{13}C_{plant} \times A_{plant})/A_{meas} \tag{2}$$

where:

A_{meas} is the measured atmospheric concentration of CO_2 in ppm;
A_{ocean} is the component of deep ocean origin, in ppm;
A_{plant} is the component of plant origin, in ppm;
$\delta^{13}C_{meas}$ is the isotopic value of measured atmospheric CO_2;
$\delta^{13}C_{ocean}$ is the isotopic value of "ocean" component atmospheric CO_2;
$\delta^{13}C_{plant}$ is the isotopic value of "plant" component atmospheric CO_2.

So, the separate components calculated using the measured values for atmospheric CO_2, and assuming isotopic values for "deep ocean" and "plant" components, are given by Equations (3) and (4) as follows.

$$A_{ocean} = A_{meas} \times (\delta^{13}C_{meas} - \delta^{13}C_{plant})/(\delta^{13}C_{ocean} - \delta^{13}C_{plant}), \tag{3}$$

and

$$A_{plant} = A_{meas} - A_{ocean}. \tag{4}$$

Measured atmospheric concentrations are recorded in parts per million (ppm) by volume. Thus, separating an ocean and a plant source component requires converting ppm by volume to ppm by mass.

So,

$$\delta^{13}C = \{R(^{13}C/^{12}C)_{meas}/R(^{13}C/^{12}C)_{std} - 1\} \times 1000 \tag{5}$$

and

$$R(^{13}C/^{12}C)_{meas} = (1 + \delta^{13}C/1000) \times R(^{13}C/^{12}C)_{std} \tag{6}$$

where:

$R(^{13}C/^{12}C)_{meas}$ is the ratio for the sample carbon-13 to carbon-12;
$R(^{13}C/^{12}C)_{std}$ is the ratio for the standard carbon-13 to carbon-12.

Then, the carbon mass may be calculated in atomic mass units as

$$\text{Carbon} = \{12 + 13 \times R(^{13}C/^{12}C)_{meas}\}/(1 + R(^{13}C/^{12}C)_{meas}). \tag{7}$$

So, the mass of CO_2 can be calculated, hence the conversion of ppm by volume to ppm by mass.

The mass correction is small, of the order of 0.0005 ppm compared with measurement errors of the order of 0.1 ppm. Although not significant, these mass corrections are applied to this data.

For the remainder of this paper, we use the term "ocean component" and "ocean source" to refer to that fraction of atmospheric CO_2 having the isotopic signature of deep ocean water.

2.2. Ocean and Plant Source Component Analysis for Monthly Values

The monthly measurements for the South Pole and Christmas Island from 1978 to 2015 and Point Barrow from 1980 to 2015 are shown in Figure 1. The ocean and plant source components of atmospheric CO_2 are found using Equations (3) and (4), respectively. The results from the monthly measurements of atmospheric CO_2 concentrations and $\delta\ ^{13}C$ are shown in Figures 2 and 3, respectively, for the South Pole and Christmas Island from 1978 to 2015 and for Point Barrow from 1983 to 2015.

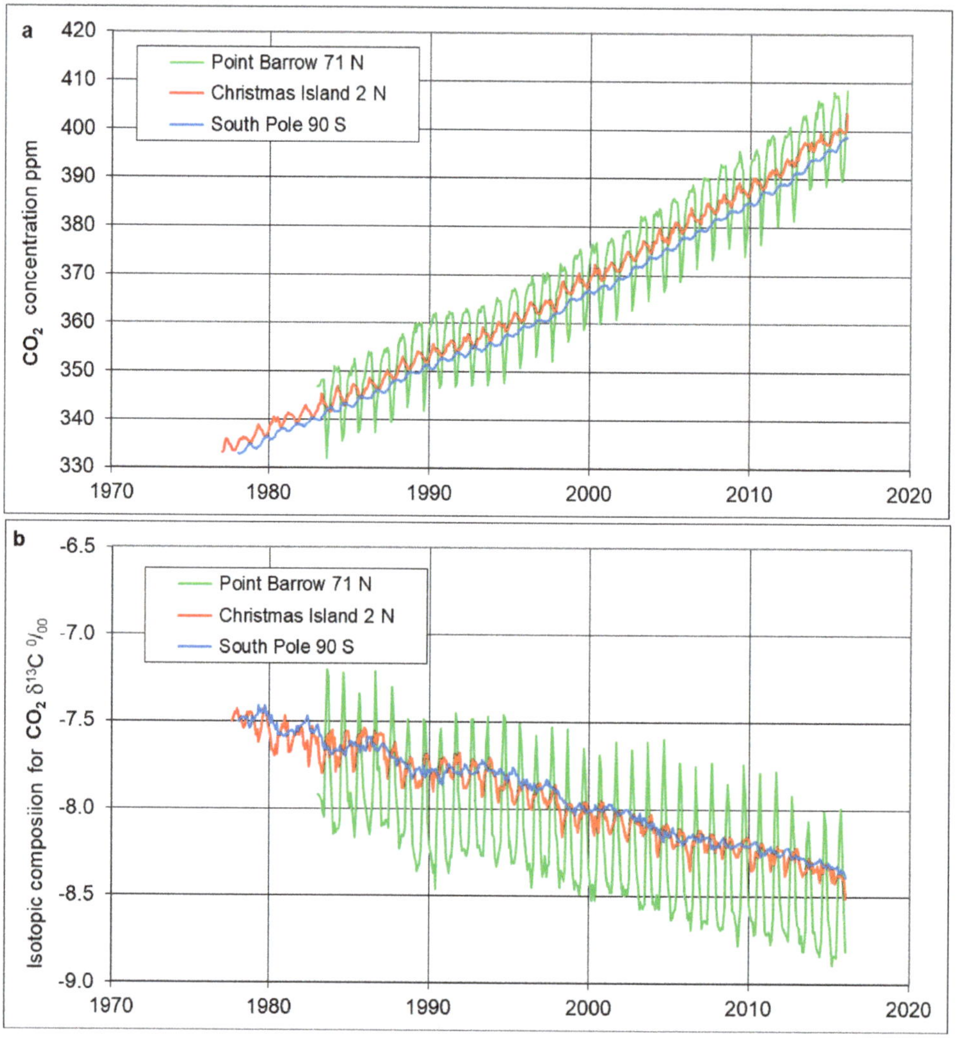

Figure 1. Monthly atmospheric CO_2 at the South Pole 90° S, Christmas Island 2° N and Point Barrow 71° N, for 1978 to 2015. (**a**) CO_2 concentration in ppm. (**b**) CO_2 isotopic composition $\delta^{13}C$ ‰.

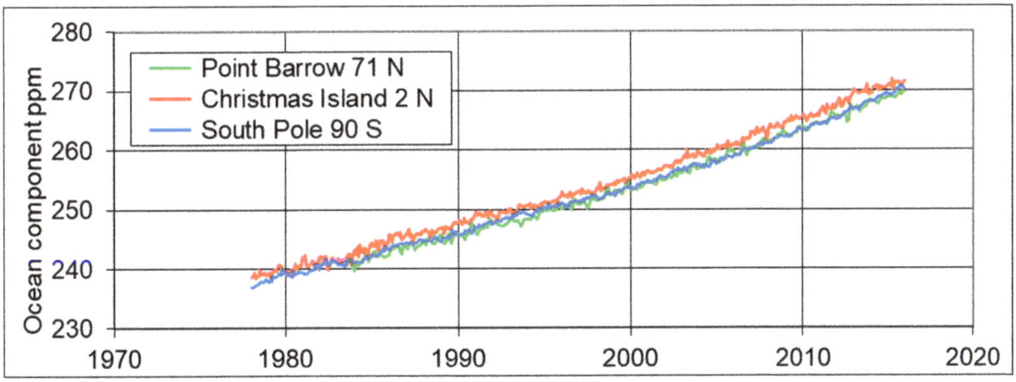

Figure 2. Monthly ocean component CO_2 concentrations at the South Pole, Christmas Island and Point Barrow for 1978 to 2015.

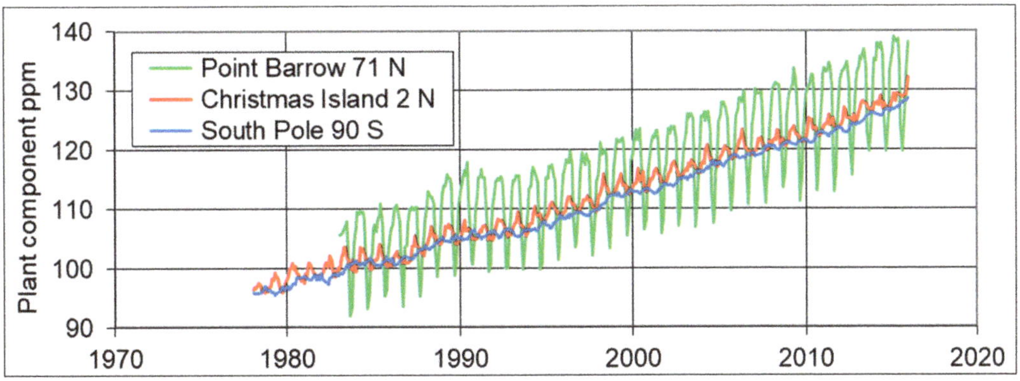

Figure 3. Monthly plant component CO_2 concentrations at the South Pole, Christmas Island and Point Barrow for 1978 to 2015. An increase is evident around 1989 (the "bubble").

The seasonal variations seen in the measured CO_2 concentrations for the South Pole, Christmas Island and Point Barrow are found in the respective plant components but not to any significant extent in the corresponding ocean components.

There are two observations that follow from this analysis:

(a) A subtle increase above the trend in the measured CO_2 concentrations is visible in Figure 1a for times around 1988–1991. This feature is clear in the plot of isotopic composition shown in Figure 1b (referred to as the "bubble" in later paragraphs) and overlaps in time the 1989 Regime Shift described by [15]. That 1989 event was clearly evident in biological records although only weakly shown in indices of the Pacific climate. Consistent with that prior finding of an event in biological records, we see that the "bubble" is present in the plant component of atmospheric CO2 as deduced from the isotopic analysis (Figure 2). However, the "bubble" is not evident in Figure 3, which shows the equivalent plot of the ocean component of atmospheric CO2. This feature is discussed in additional detail below in Section 4.

(b) For the ocean components shown in Figure 2, Christmas Island (2° N) near the equator shows higher values for the entire time span of 1978–2015, while Point Barrow (71° N) in the Arctic shows the lowest values. We discuss the significance of these observations in the context of ocean currents, in Section 3.2.

The average annual increases for each of the stations in Figure 1, computed for the time span from 1986 to 2015, are found through a least-squares fit to the measurements; these are listed in Table 2. The most remarkable result is that, for each site, the ocean and plant average annual increases are similar, to within 0.3%.

Table 2. Annual CO_2 increases using monthly data, averaged over the years of 1986 to 2015.

		Annual Increases in ppm	
Stations	**CO_2 Concentration**	**Ocean Component**	**Plant Component**
South Pole 90° S	1.769 +/− 0.008	0.886 +/− 0.005	0.883 +/− 0.005
Christmas Island 2° N	1.819 +/− 0.009	0.910 +/− 0.005	0.909 +/− 0.007
Point Barrow 71° N	1.800 +/− 0.038	0.902 +/− 0.005	0.898 +/− 0.038

3. Ocean and Plant Source Component Latitude Analysis

3.1. Annual Values

The behavior of the ocean and plant components are further explored through analysis of each of the ten SIO latitudes listed in Table 1.

The variation in atmospheric CO_2 measurements shows increasing concentrations from the lowest values at the South Pole 90° S to a peak at Point Barrow 71° N. Figure 4 shows the residual annual variations of total CO_2 less the value at the South Pole, for each station, averaged over 1986 to 2015. This is the time span where there are concentration and isotopic composition CO_2 measurements available for all latitudes shown in Table 1.

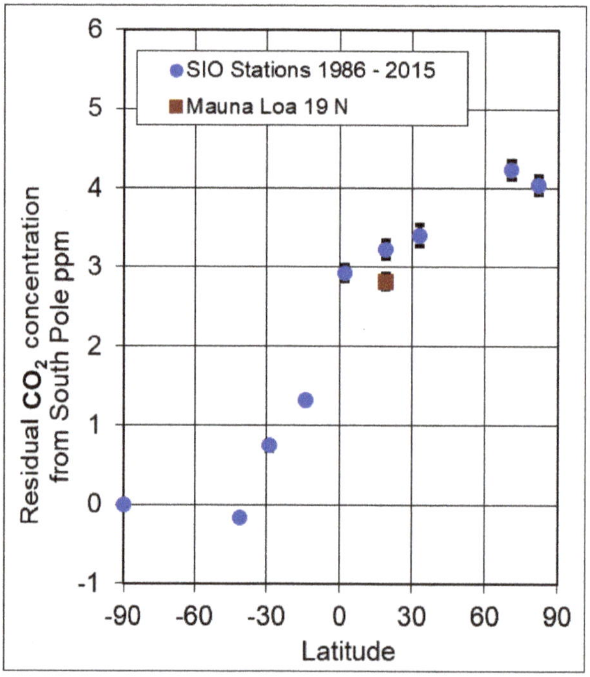

Figure 4. Residual annual variations in total CO_2 less the value at the South Pole, averaged for 1986 to 2015, for each of the ten stations listed in Table 1. Black vertical bars show standard deviations associated with blue data points.

The latitude variations in the residual annual variations of total CO_2 concentrations are quite different when separated into ocean and plant components. The component values, standard deviations and errors of the mean are listed in Tables 3 and 4 and shown in Figure 5. The ocean component peaks near the equator (Figure 5a). However, the plant component peaks at the Arctic Circle latitudes of Point Barrow 71° N and Alert 82° N (Figure 5b).

Table 3. Residual ocean component values for 1986 to 2015.

SIO Stations	Latitude	Longitude	Elevation (Meters)	Residual Values from South Pole Value in ppm		
				Mean	Standard Deviation	Error of Mean
Alert	82° N	63° W	210	−0.29	0.33	0.06
Point Barrow	71° N	157° W	11	−0.30	0.39	0.06
La Jolla	33° N	117° W	10	0.36	0.40	0.07
Kumukahi	19° N	155° W	3	0.70	0.39	0.07
Mauna Loa	19° N	156° W	3397	0.99	0.24	0.04
Christmas Island	2° N	157° W	2	1.62	0.42	0.08
American Samoa	14° S	171° W	30	0.94	0.26	0.05
Kermadec	29° S	178° W	2	0.43	0.34	0.06
Baring Head	41° S	175° E	85	0.16	0.27	0.05
South Pole	90° S		2810	0.00		

Table 4. Residual plant component values for 1986 to 2015.

SIO Stations	Latitude	Longitude	Elevation (Meters)	Residual Values from South Pole Value in ppm		
				Mean	Standard Deviation	Error of Mean
Alert	82° N	63° W	210	4.33	0.57	0.10
Point Barrow	71° N	157° W	11	4.54	0.58	0.11
La Jolla	33° N	117° W	10	3.04	0.74	0.14
Kumukahi	19° N	155° W	3	2.52	0.52	0.09
Mauna Loa	19° N	156° W	3397	1.83	0.59	0.11
Christmas Island	2° N	157° W	2	1.31	0.48	0.09
American Samoa	14° S	171° W	30	0.37	0.30	0.05
Kermadec	29° S	178° W	2	0.31	0.33	0.06
Baring Head	41° S	175° E	85	−0.32	0.24	0.04
South Pole	90° S		2810	0.00		

The analysis also shows significantly different residual values at Mauna Loa at an elevation of 3397 m compared to Kumukahi at the same latitude but at an elevation of 3 m (highlighted in Tables 3 and 4). The ocean component residual at Mauna Loa is significantly greater than the Kumukahi residual and the sign of this difference of residuals is reversed for the corresponding plant components. These differences are shown in Figure 5.

The differences may be due to the elevation at which Mauna Loa measurements are made; we suggest that the elevation of 3397 m is above the sub-tropical inversion, having the effect that the atmosphere is locally more mixed with contributions from the equatorial region. This would have the effect of increasing the residual ocean component relative to that found at Kumukahi at an elevation of 3 m. Likewise, the residual plant component values at the elevated Mauna Loa site are mixed to a lower value than the Kumukahi residual plant value.

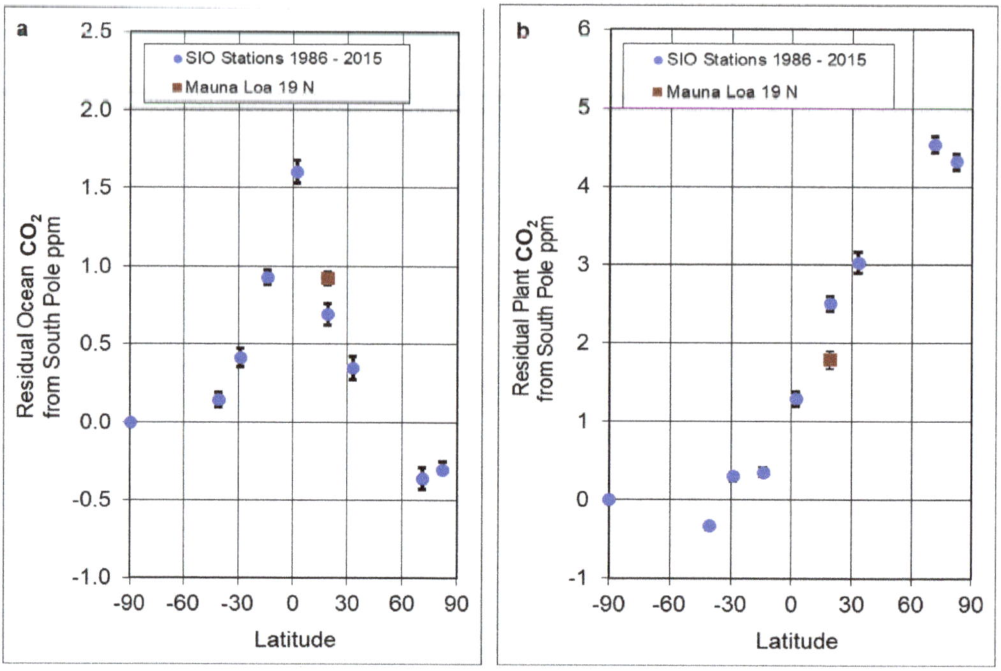

Figure 5. Residual annual variations in CO_2 source components, less the value at the South Pole, averaged for 1986 to 2015. (**a**) The ocean component and (**b**) the plant component.

We also note from Figures 5 and 6 that the negative values of CO_2 relative concentration, and of plant source component CO_2, as observed at Baring Head (latitude 41° S between the north and south islands of New Zealand), are associated with a location between the Southern Ocean and south Pacific Ocean. This is consistent with an understanding that the expanse of ocean in the southern hemisphere acts as a sink for atmospheric CO_2, and further, that an absence of upwelling deep ocean currents in southern latitudes [9] is associated with an absence of an anomaly in the ocean source component of CO_2 at Baring Head.

3.2. *Latitude Analysis for Seasonal Values*

If the ocean source component is independent of the plant source component, then the ocean source peaking at the equator should be independent of seasonal variations. Figure 6 shows residual seasonal variations restricted to the end of the boreal winter and to the end of the boreal summer. Averaging is performed for the two months of February–March, and for August–September. As for Figure 5, we obtain ocean and plant components of CO_2 for all latitudes, but restricted to the two seasons. There is little change in the ocean components for the seasonally separated values relative to the residual annual averages. The stability of the ocean source CO_2 concentrations can be seen by comparing the latitude seasonal changes for February–March with those of August–September (Figure 6a,b).

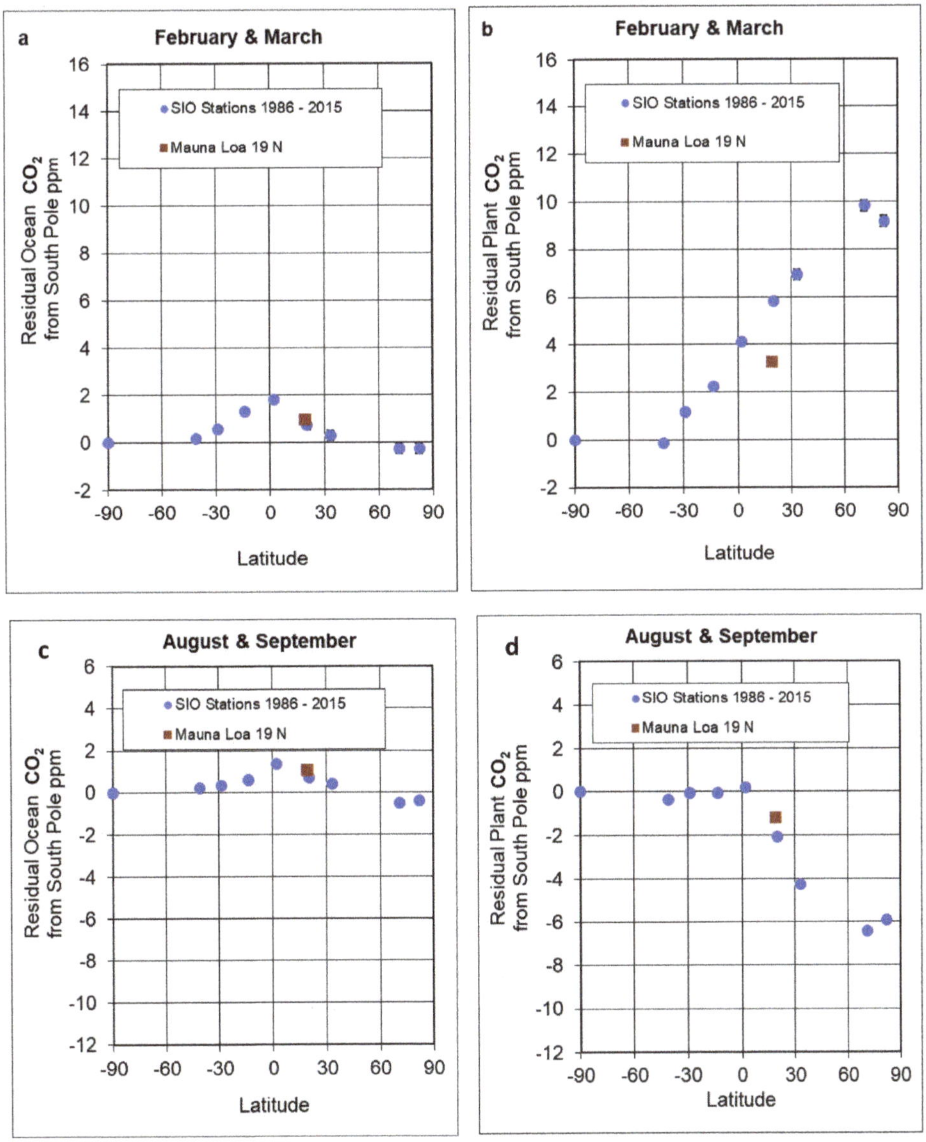

Figure 6. Residual seasonal variations in CO_2 source components, less the value at the South Pole, averaged for 1986 to 2015. (**a**) Ocean source for end boreal winter, average of February and March (**b**) plant source for end boreal winter, (**c**) ocean source for end boreal summer, average of August and September and (**d**) plant source for end boreal summer.

The residual seasonal variations in the plant component of CO_2 are quite different when compared to the ocean component, as seen in plots of Figure 6b,d. Plant components show changes of up to 16 ppm from the end of the boreal winter (February–March) to end of summer (August—September) at the far-north stations of Point Barrow and Alert.

The hypothesis for this seasonal difference is that plant source component variations in the Northern Hemisphere are due to the influence on the carbon cycle of the boreal forests. The trees start to grow in spring, drawing CO_2 from the atmosphere until autumn when the end of summer growth reduces the draw-down of CO_2. In addition, it is expected that decaying plants will return CO_2 to the atmosphere.

The seasonal differences of end boreal winter less boreal summer are shown in Figure 7 for ocean and plant components. The ocean component (Figure 7a) shows that seasonal differences are not statistically different with the exception of Christmas Island 2° N, American Samoa 14° S and Kermadec Island 29° S. The peak in the ocean component for the boreal winter (Figure 6a) is shifted about 15° S in Figure 7a. This may be explained as a result of transport by atmospheric winds comparatively enriched in the ocean component as there are strong wind systems, particularly the trade winds in the equatorial region.

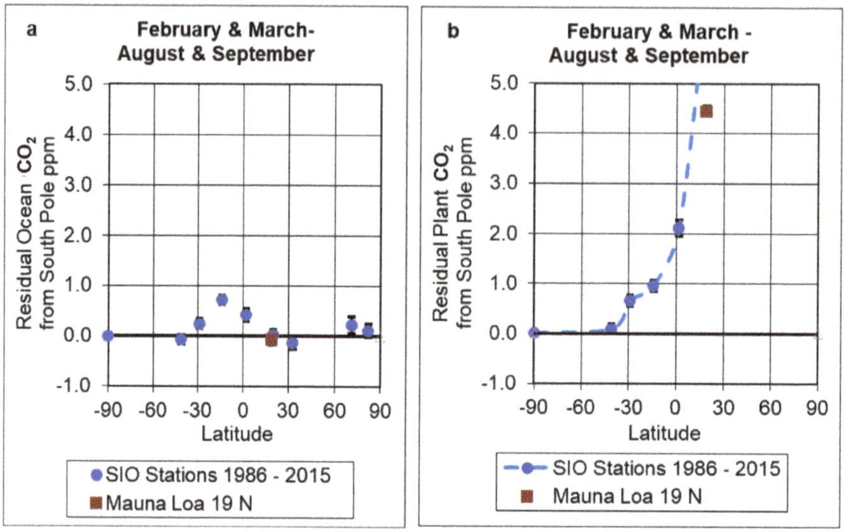

Figure 7. Seasonal differences for end boreal winter less end boreal summer. Black vertical bars show standard deviations associated with blue data points. (**a**) The residual ocean component and (**b**) the residual plant component.

The plant component also shows evidence of atmospheric transport from the northern hemisphere to southern hemisphere during the boreal winter, seen at latitudes 15° S to 30° S (Figures 6b,d and 7b). It is seen most clearly in Figure 7b at 30° S where the residual plant component is approximately 0.5 ppm above the trend, consistent with atmospheric transport from north to south.

This analysis shows that at Point Barrow, 71° N, and at Alert, 82° N, the ocean component varies only slightly by 0.3 ± 0.2 ppm and 0.1 ± 0.1 ppm from boreal winter to summer, whereas there are large variations of 16 ppm and 15 ppm for the plant component, respectively. At lower latitudes in the northern hemisphere, there are reduced differences from winter to summer. The spatial and seasonal distribution in ocean and plant source components differs by up to a factor of 50, and leads us to the conclusion that the ocean source of CO_2 is an independent source of atmospheric CO_2 and is decoupled from the plant source component.

3.3. Sensitivity of Latitude Analysis to Assumptions of Isotopic Composition

The separation of observations of atmospheric CO_2 content into ocean and plant source components is made using assumptions that the CO_2 ocean component $\delta^{13}C$ is 0‰ and the plant component $\delta^{13}C$ is −26‰. The sensitivity to this choice is assessed by independently varying each component $\delta^{13}C$ by ±2. The results are shown in Figure 8 for station variations less the source contributions at the South Pole, for the time span of 1986 to 2015.

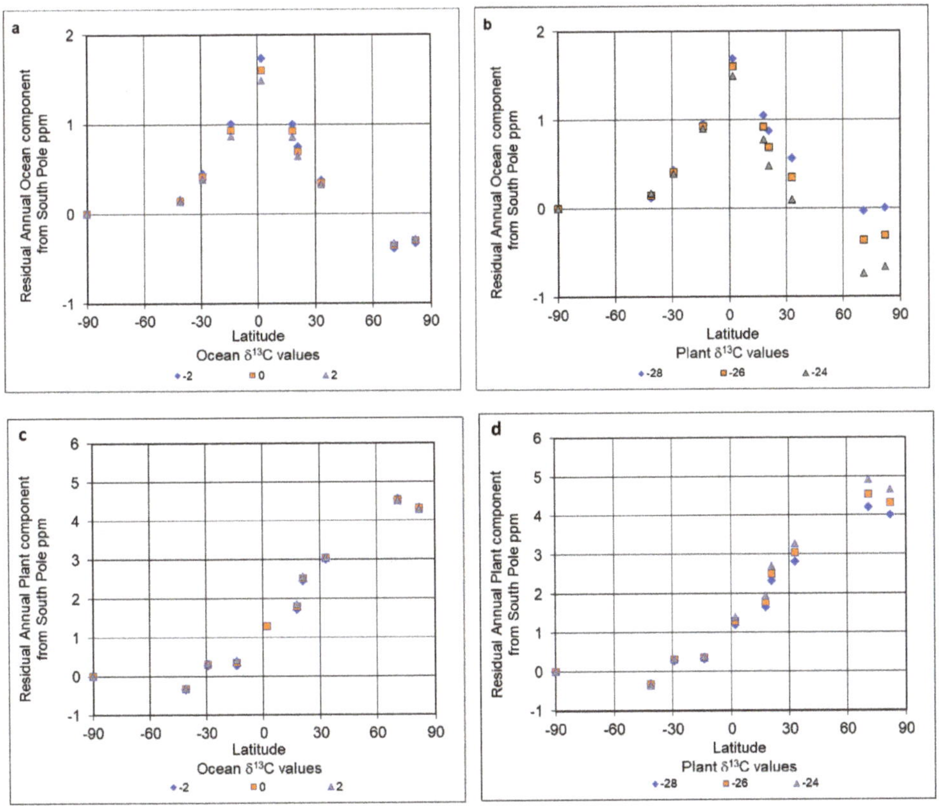

Figure 8. Sensitivity of computed residual annual variations in CO_2 source components, less the value at the South Pole, averaged for 1986 to 2015. (**a**) Residual ocean component using $\delta^{13}C_{ocean}$ = 0‰ (as for Figure 5a), plus computed values where $\delta^{13}C_{ocean}$ is perturbed by ±2. (**b**) Residual ocean component using $\delta^{13}C_{plant}$ = −26‰, and also perturbed by ±2. (**c**) Residual plant component using $\delta^{13}C_{ocean}$ = 0 (as for Figure 5b), plus computed values where the $\delta^{13}C_{ocean}$ value of 0‰ is perturbed by ±2. (**d**) Residual plant component using the value $\delta^{13}C_{plant}$ = −26‰, and also perturbed by ±2.

The perturbed values of $\delta^{13}C_{ocean}$ in Figure 8 give a maximum component variation of 0.1 ppm, while the corresponding perturbations in $\delta^{13}C_{plant}$ give a maximum component variation of 0.4 ppm.

These perturbation studies demonstrate that the patterns of ocean and plant component changes with latitude are not significantly affected by variations of ±2‰ in the assumed baseline values used for $\delta^{13}C$.

4. Annual Differences of Ocean and Plant CO_2 Components

4.1. Annual Differences for Individual Stations

Figures 1–3 show plots of monthly variations in total CO_2 and its isotopically separated ocean and plant components. Additional features of ocean and plant source components of CO_2 can be seen in an analysis of annually averaged measured CO_2 concentrations and corresponding ocean and plant components. These are shown in Figures 9–11. Figures 10 and 11 show that there is more annually averaged variability in the plant components compared to the ocean components. The simplest demonstration of variability comes from the standard deviations for the year-on-year changes. The results, given in Table 5, show that the ocean components are consistently less variable than the plant components.

We suggest that this may be attributable to the fact that the ocean dynamics change more slowly than the atmospheric dynamics, affecting plant components.

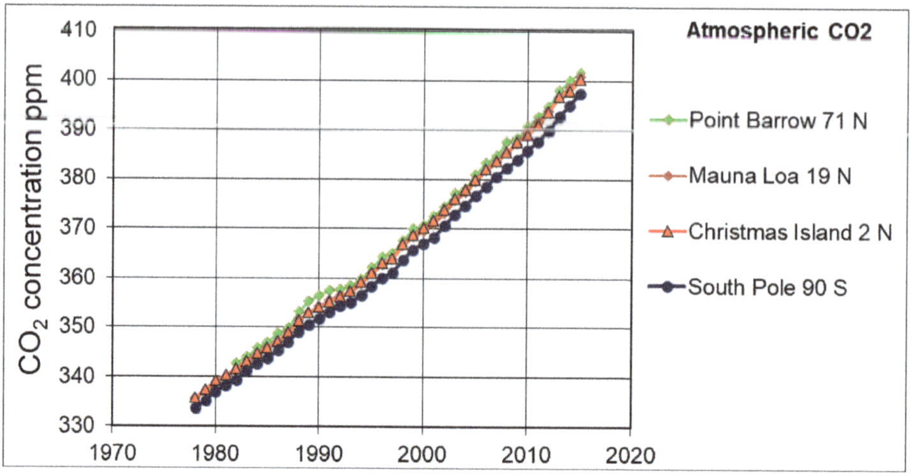

Figure 9. Annually averaged CO_2 measurements for the South Pole, Christmas Island, Mauna Loa and Point Barrow.

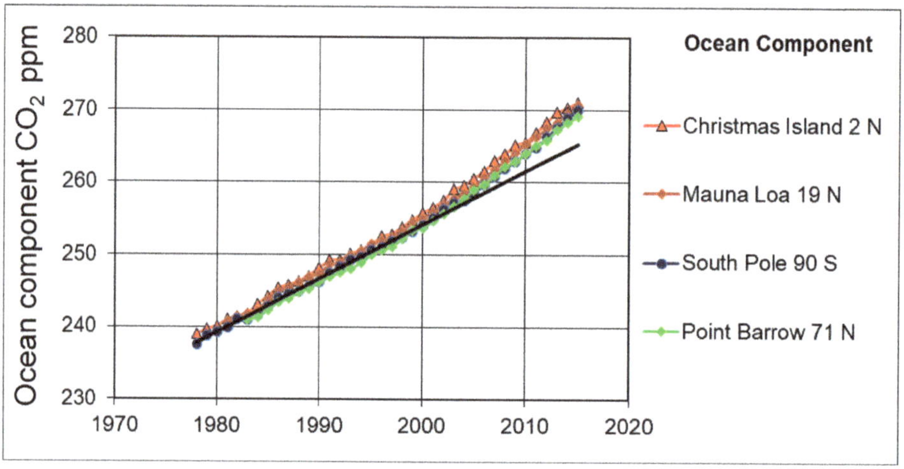

Figure 10. Ocean component values obtained from annually averaged CO_2 observations for the South Pole, Christmas Island, Mauna Loa and Point Barrow. Note the straight-line fit (black line) for the South Pole from 1979 to 2000, which is projected to 2015.

The quantitative annual increases for atmospheric CO_2 and components from 1986 to 2015 are given in Table 6. The ocean component and plant components are approximately equal at 50% of the annual CO_2 concentration increase.

We note that the annual values in Table 2 are directly from monthly values that have seasonal variations, while the values in Table 6 come from annual averages; hence, the latter do not show seasonal variations. This difference is illustrated in a comparison of the plant component at Point Barrow, which has a larger standard deviation of 0.038 ppm from the monthly seasonal variations, compared with 0.02 ppm for the annual averages.

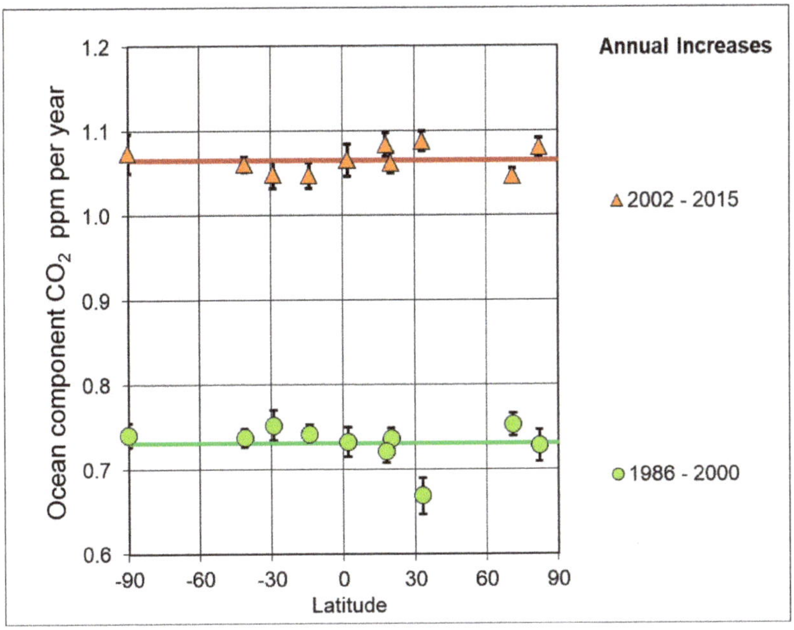

Figure 11. Annual increases in ocean component CO_2, averaged for all stations. The increase is 0.73 ppm per year for 1986 to 2000 and 1.07 ppm per year for 2002 to 2015.

Table 5. Station variability, measured as standard deviations of annually averaged CO_2, ocean and plant components.

	Year-on-Year Changes: Standard Deviations in ppm		
Stations	**CO_2 Concentration**	**Ocean Component**	**Plant Component**
South Pole 90° S	0.46 +/− 0.08	0.31 +/− 0.06	0.48 +/− 0.09
Christmas Island 2° N	0.57 +/− 0.10	0.38 +/− 0.07	0.53 +/− 0.10
Mauna Loa 19° N	0.54 +/− 0.10	0.32 +/− 0.06	0.59 +/− 0.10
Point Barrow 71° N	0.86 +/− 0.14	0.28 +/− 0.05	0.74 +/− 0.14

Table 6. Annual CO_2 increases using annual data, averaged over the years of 1986 to 2015.

	Annual Increases in ppm		
Stations	**CO_2 Concentration**	**Ocean Component**	**Plant Component**
South Pole 90° S	1.77 +/− 0.03	0.89 +/− 0.02	0.88 +/− 0.02
Christmas Island 2° N	1.82 +/− 0.03	0.91 +/− 0.02	0.91 +/− 0.02
Mauna Loa 19° N	1.82 +/− 0.03	0.88 +/− 0.02	0.94 +/− 0.02
Point Barrow 71° N	1.81 +/− 0.03	0.90 +/− 0.02	0.91 +/− 0.02

4.2. Trend Breaks in Annual Differences for the Ocean Component

The annually averaged CO_2 concentration values are shown in Figure 9. It is obvious that there is a slight excursion from 1988 to 1991 for Point Barrow CO_2 concentrations (the "bubble" referred to above).

For the ocean component shown in Figure 10, the Christmas Island values are above those at the South Pole and Point Barrow. Further, a straight-line fit for the South Pole component from 1979 to 2001 projected to 2018 shows a trend change in ocean component data at about year 2000.

We use the Chow Break Test [16] on each of the curves plotted in Figure 10. The Chow Break Test is used on a data series, and it uses a linear least-squares straight-line fit to show whether there is a statistically significant break for two straight lines to be a better fit than one straight line. This example shows that the ocean component for each of the South Pole, Christmas Island, Mauna Loa and Point Barrow has a break point in its trend at year 2001 ± 2.

This break in the ocean component is present at all latitudes, as shown in Figure 11, where the average ocean component of CO_2 increases by 0.731 ± 0.008 ppm per year for 1986 to 2000 and 1.066 ± 0.005 per year for 2002 to 2015.

In searching for an explanation, or at least an association, with the break in the ocean component, we observe that the break in ocean component CO_2 after 2001 is coincident with an inflection in global lower troposphere temperatures measured by satellites [17]. The inflection is obvious in Figure 12 where separate straight-line fits for 1979 to 2000 and 2001 to 2015 are identified. Tropospheric temperatures show a rising trend of the order of 0.15 °C per decade before 2001, and a temperature plateau starting in 2000 to 2002, extending to approximately 2015 (the duration of this study).

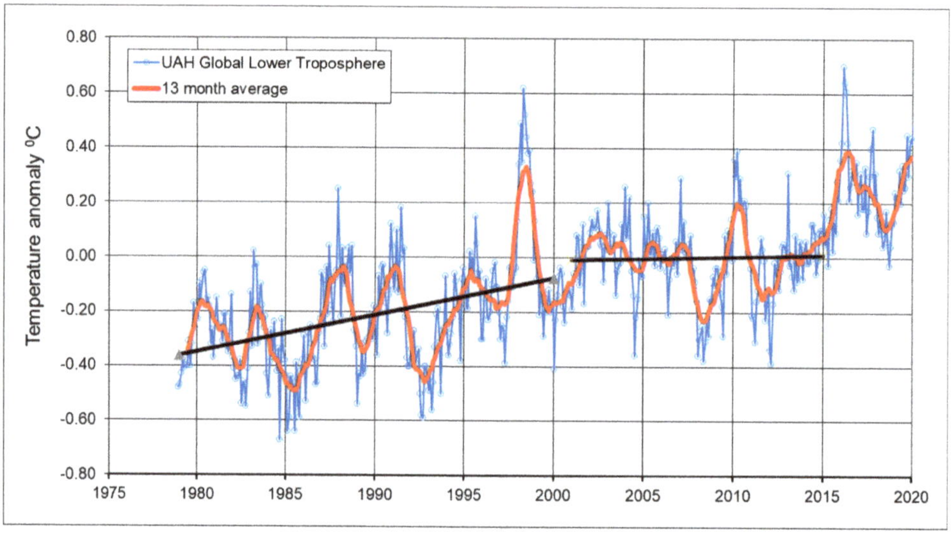

Figure 12. Global lower troposphere anomaly from UAH satellite measurements [17]. Solid straight lines are least-squares fits for 1979 to 2000 and 2001 to 2015.

An analysis of the global carbon cycle by [4] also shows a change in trend following the year of 2000. That analysis used measured and modeled data (but not isotopic data) to estimate the movement of CO_2 from the atmosphere into the ocean and found an increase in CO_2 absorbed into the ocean from about 2000 to 2019. In contrast to [4], the results in Figures 11 and 12 demonstrate a change in the trend of movement of CO_2 from ocean to atmosphere, which is based on observational data only.

A hypothesis consistent with the observed break in the trends of atmospheric ocean-component CO_2 and tropospheric temperature around 2001 is that there is an increased upwelling of cold water in the oceans, which may then reduce global temperature and release CO_2 with an increased ocean-component isotopic mix into the atmosphere.

4.3. Annual Differences in the Plant Component CO_2; Correlation with a Phase Change in the AMO

The plant component CO_2 also shows an observed temporal change associated with ocean and atmospheric temperature shifts. Figure 13 shows annual changes including the "bubble" for the years 1988–1992 associated with the 1989 Regime Shift described by [15]. Replotting these data in Figure 14 after trend removal shows not only the "bubble" but a correlation of plant component CO_2 with increasing latitude—a regional rather than global change. The residual plant CO_2 values show increases in a northerly direction from the South Pole to Point Barrow 71° N. The obvious hypothesis to explain this variation is that the source of atmospheric plant CO_2 is a combination of fossil fuel CO_2, phytoplankton productivity in the oceans, and forest and peat fires, which occur primarily in the northern hemisphere. Figure 14 is consistent with this combination of sources.

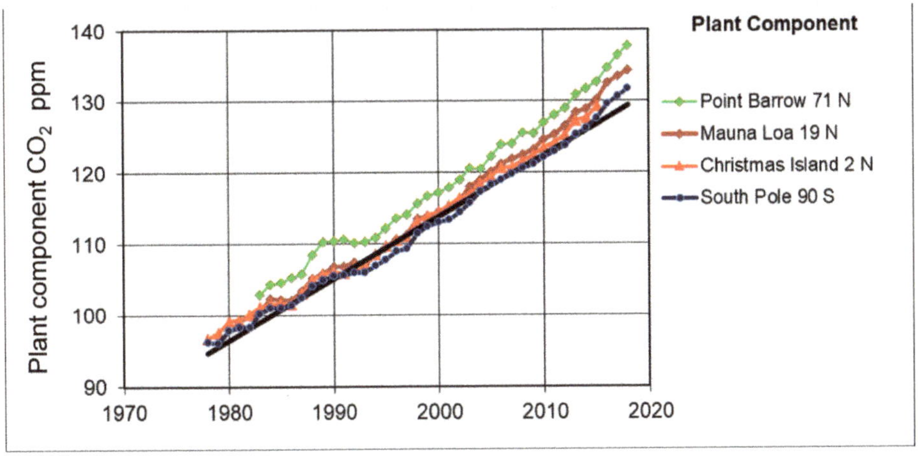

Figure 13. Annual plant component CO_2 for the South Pole, Christmas Island, Mauna Loa and Point Barrow. The 1989 regime shift ("bubble") in plant component atmospheric CO_2 is most apparent for the northernmost latitude, and for the years 1988 to 1992. The black straight line is a least-squares fit to the plant component CO_2 from 1978 to 2018 at the South Pole.

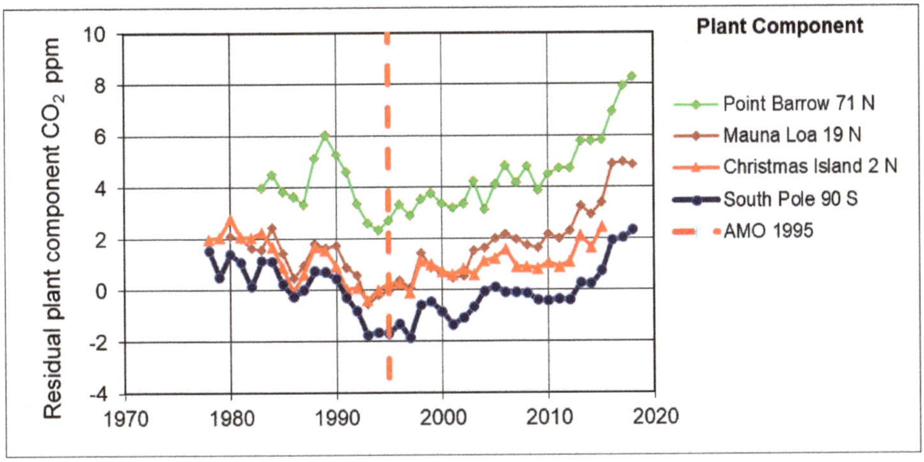

Figure 14. Residual annual values from the straight-line fit to the South Pole measurements shown in Figure 13. The vertical dashed line at 1995 denotes the time of the phase change in the AMO as given by [18]. Note also the presence of the "bubble" in the years of 1988 to 1992.

Figure 14 also shows a clear inflection in the residual trends of the annual change in plant component CO_2 near the year of 1995. This coincides with a phase change in the Atlantic Multi-decadal Oscillation (AMO) from negative to positive, as described by [18]. That phase change is also associated with synchronous changes in numbers of small pelagic fishes [19]. These observational changes are consistent with an interpretation of decreased phytoplankton productivity after 1995, which in turn may explain the observed decrease in numbers of small pelagic fishes.

The pelagic fish time series shows an increase from 1995 to 2010 and, with phytoplankton at the bottom of the ocean food chain, implies increasing phytoplankton productivity. Such an increase would cause the increased removal of CO_2 in the ocean; this outcome in turn is associated with a decrease in the plant component of atmospheric CO_2. Thus, variable ocean sinks for CO_2 are directly coupled to the atmosphere as the concentration and isotopic composition of the atmosphere are varied.

4.4. Correlation of Global Plant Component CO_2 Variations with the ENSO

It is possible to estimate the annual changes in global atmospheric CO_2 concentrations and the separate ocean and plant CO_2 components, using the nine surface stations listed in Table 1 (where Mauna Loa is excluded, being at an altitude of 3397 m above sea level). The nine stations are taken to represent values in latitude bands defined by the average of the station latitudes, with the exception of the latitude extremes represented by Alert and the South Pole. The band latitudes are then used to calculate the fraction of the global surface within the bands, as shown in Table 7.

Table 7. Latitude bands for global CO_2 concentrations.

Station	Latitude	Latitude Band	Band width Degrees	Fraction of Global Surface
Alert	82° N	90° N to 76.5° N	13.5	1.38%
Point Barrow	71° N	76.5° N to 52° N	24.5	9.22%
La Jolla	33° N	52° N to 26° N	26.0	17.48%
Kumukahi	19° N	26° N to 10.5° N	15.5	12.81%
Christmas Island	2° N	10.5° N to 6° S	16.5	14.34%
American Samoa	14° S	6° S to 21.5° S	15.5	13.10%
Kermadec Island	29° S	21.5° S to 35° S	13.5	10.35%
Baring Head	41° S	35° S to 65.5° S	30.5	16.82%
South Pole	90° S	65.5° S to 90° S	24.5	4.50%
Total			**180.0**	**100.00%**

Figure 15a–c show the year-on-year annual changes in global atmospheric CO_2 concentrations and annual changes in global plant and ocean CO_2 components. For comparison, the El Nino southern Oscillation (ENSO) variation is shown in Figure 15d via the Nino 3.4 SST Index [2,20]. The strong 1997–1998 and 2016–2017 El Nino events are obvious. Between 1989 and 2013, there are 10 El Nino peaks, which correlate in time with peaks in the annual plant component of CO_2, although the two time series are clearly not linearly related in their magnitude.

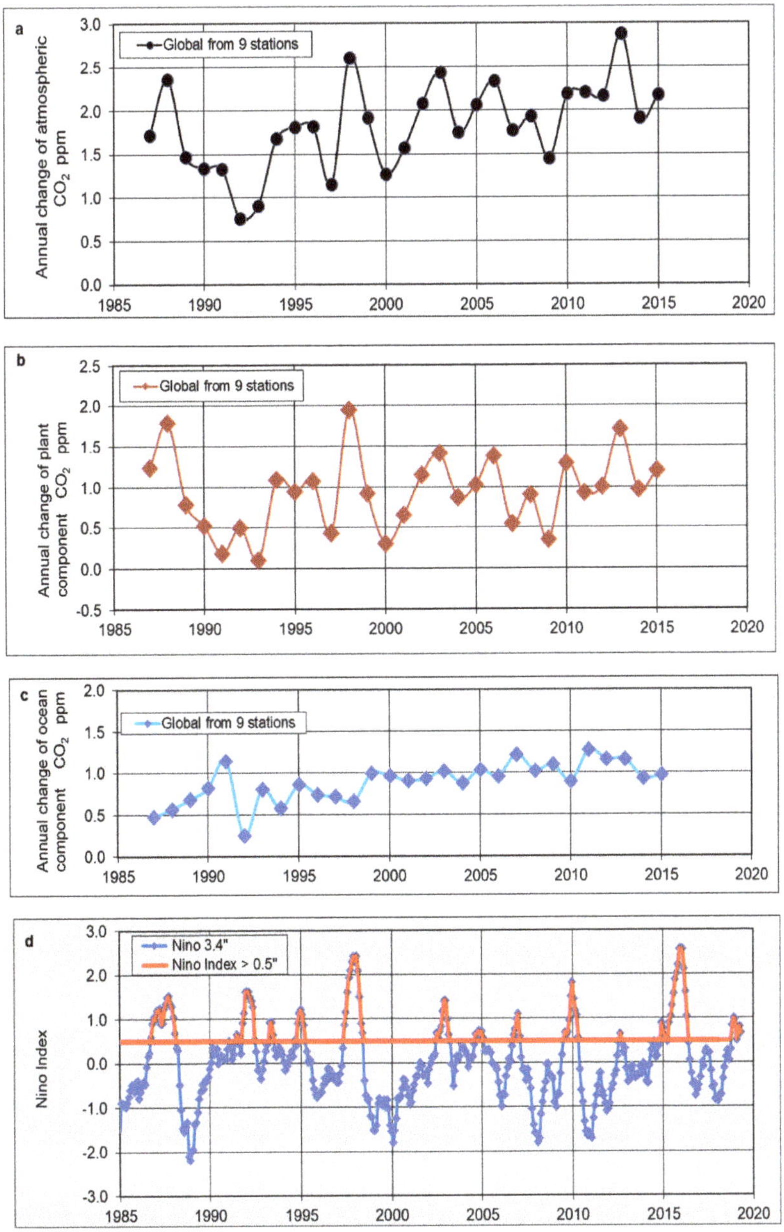

Figure 15. Annual changes in atmospheric CO_2 concentrations with ocean and plant components of CO_2 in the atmosphere, averaged for 9 stations listed in Table 7. (**a**) Atmospheric CO_2 concentrations, (**b**) plant component CO_2, (**c**) ocean component CO_2 and (**d**) Nino 3.4 Index. Red lines correspond with El Nino conditions (index > 0.5).

5. Discussion and Conclusions

For the time span of 1978 to 2015, the division of atmospheric CO_2 concentrations into two components having separate origins from deep ocean sources and from plants including fossil fuel emissions, shows a differing geographical distribution of the components. This indicates that variations in deep ocean components and in plant components of atmospheric CO_2 concentrations are attributable to independent mechanisms. These results are not significantly affected by variations in the assumed baseline values of $\delta^{13}C$, which characterize the isotopic composition of carbon in CO_2.

We find that the ocean component and plant components as derived from isotopic data are approximately equal at 50% of the annual CO_2 concentration increase. This result is similar to the outcome of a recent study [21] that used an ocean circulation model (without isotopic data) to track emitted carbon through the atmosphere–ocean system from pre-industrial times to the present (1780 to 2020). That study provided an estimate of 55% ocean source in atmospheric CO_2 despite using a very different methodology compared with this paper. It is of interest that the two different approaches yield estimates of those contributions to atmospheric CO_2 concentrations that are similar within a few percentage points.

Four essential conclusions may be drawn from this study using carbon isotopes.

1. Annual increases in atmospheric CO_2 concentrations over 1978–2015 when split into plant and deep ocean sources via the isotopic signatures show approximately equal parts when averaged over all latitudes.
2. The deep ocean component of CO_2 has its maximum value at the equator and we conclude that it is an independent source of atmospheric CO_2. This is further demonstrated by its limited year-on-year variability. We hypothesize that the observations of changes in ocean source CO_2 are attributable to the upwelling of CO_2-rich water transported across the sea floor via the global thermohaline circulation of the oceans.
3. The plant source CO_2 in the atmosphere shows strong seasonal variability in northern latitudes up to the Arctic circle. It also shows more temporal year-by-year variability than that of the ocean source CO_2 and shows maximum variations coincident with the timing of El Nino events. It also shows a change in trend associated with the 1995 AMO phase change. These spatial and temporal variations are consistent with the plant source CO_2 being associated with multiple variations in fossil fuel CO_2 emissions, together with natural variations in the net primary productivity of oceanic phytoplankton, and forest and peat fires.
4. The differing variations by year and by latitude of plant source and deep ocean source CO_2 as measured from the analysis of carbon isotopes in atmospheric CO2 samples, suggest that inclusion of such isotopic data in full carbon cycle models for the Earth may provide additional constraints on the modeling of bidirectional atmosphere–ocean transfers of CO_2.

Author Contributions: Conceptualization and software, T.Q.; methodology, T.Q. and M.A.; validation, formal analysis, investigation, resources, data curation, T.Q.; writing—original draft preparation, T.Q.; writing—review and editing, T.Q. and M.A. All authors have read and agreed to the published version of the manuscript.

Funding: This research received no external funding.

Institutional Review Board Statement: Not applicable.

Informed Consent Statement: Not applicable.

Data Availability Statement: Scripps Institution of Oceanography (SIO) [6] CO_2 and $\delta^{13}C$ isotope variations of CO_2 summarized in Table 1 are available at https://scrippsco2.ucsd.edu/data/atmospheric_co2/sampling_stations.html, (accessed 21 June 2023). For each observation station, we use data downloaded in files "monthly_flask_co2_alt.csv" and "monthly_flask_c13_alt.csv", where we use column 9 (filled data interpolating over gaps in the monthly record). Global lower troposphere temperatures measured by satellites (Earth System Science Center, 2022) used in Figure 14 were accessed from columns 1, 2, 3 of https://www.nsstc.uah.edu/data/msu/v6.0/tlt/uahncdc_lt_6.0.txt (accessed 21 June 2023). The Nino 3.4 SST Index used in Figure 15d was accessed from NOAA [2] at https://psl.noaa.gov/gcos_wgsp/Timeseries/Data/nino34.long.data (accessed 21 June 2023).

Conflicts of Interest: The authors declare no conflict of interest. Author Tom Quirk is a member and former board member of the Australian Environment Foundation. The Foundation had no role in the design of the study; in the collection, analyses, or interpretation of data; in the writing of the manuscript; or in the decision to publish the results. The paper reflects the views of the authors and not the Foundation. Author Michael Asten is an associate of Earth Insight. The paper reflects the views of the scientists and not Earth Insight.

References

1. Quirk, T. Did the global temperature trend change at the end of the 1990s? *Asia-Pac. J. Atmos. Sci.* **2012**, *48*, 339–344. [CrossRef]
2. NOAA. Nino 3.4 SST Index. Available online: https://psl.noaa.gov/gcos_wgsp/Timeseries/Data/nino34.long.data (accessed on 15 April 2023).
3. Crisp, D.; Dolman, H.; Tanhua, T.; McKinley, G.A.; Hauck, J.; Bastos, A.; Sitch, S.; Eggleston, S.; Aich, V. How well do we understand the land-ocean-atmosphere carbon cycle? *Rev. Geophys.* **2022**, *60*, e2021RG000736. [CrossRef]
4. Gruber, N.; Bakker, D.C.E.; DeVries, T.; Gregor, L.; Hauck, J.; Landschützer, P.; McKinley, G.A.; Muller, J.D. Trends and variability in the ocean carbon sink. *Nat. Rev. Earth Environ.* **2023**, *4*, 119–134. [CrossRef]
5. Machado, K.S.; Froehner, S. Changes in atmospheric CO2 levels recorded by the isotopic signature of n-alkanes from plants. *Glob. Planet. Chang.* **2017**, *148*, 72–78. [CrossRef]
6. Keeling, C.D.; Piper, S.C.; Bacastow, R.B.; Wahlen, M.; Whorf, T.P.; Heimann, M.; Meijer, H.A. Exchanges of Atmospheric CO_2 and $13CO_2$ with the Terrestrial Biosphere and Oceans from 1978 to 2000. In *Global Aspects*; SIO Reference Series; No. 01-06; Scripps Institution of Oceanography: San Diego, CA, USA, 2001; 88p. Available online: https://scrippsco2.ucsd.edu/data/atmospheric_co2/sampling_stations.html (accessed on 21 June 2023).
7. Hoffman, D.W.; Rasmussen, C. Absolute carbon stable isotope ratio in the Vienna Peedee Belemnite isotope reference determined by ^{1}H NMR spectroscopy. *Anal. Chem.* **2022**, *94*, 5240–5247. [CrossRef] [PubMed]
8. Kroopnick, P.M. The distribution of ^{13}C of ΣCO2 in the world oceans. *Deep-Sea Res.* **1985**, *32*, 57–84. [CrossRef]
9. Ohmoto, H. Stable isotope geochemistry of ore deposits. *Rev. Mineral.* **1986**, *16*, 491–559.
10. Kozyr, A. *Carbon Dioxide, Hydrographic, and Chemical Data Obtained during the R/V Knorr Repeat Hydrography Cruise in the Atlantic Ocean: CLIVAR CO2 Sections A20_2003 (22 September–20 October 2003) and A22_2003 (23 October–13 November, 2003)*; ORNL/CDIAC-154; NDP-089; Carbon Dioxide Information Analysis Center, Oak Ridge National Laboratory, U.S. Department of Energy: Oak Ridge, TN, USA, 2008.
11. NOAA. The Global Conveyor Belt. Available online: https://oceanservice.noaa.gov/education/tutorial_currents/05conveyor2.html (accessed on 21 June 2023).
12. O'Leary, M. Carbon isotopes in photosynthesis fractionation techniques may reveal new aspects of carbon dynamics in plants. *BioScience* **1988**, *38*, 328–336.
13. Goericke, R.; Fry, B. Variations of marine plankton fi13C with latitude, temperature, and dissolved CO_2 in the world ocean. *Glob. Biogeochem. Cycles* **1994**, *8*, 85–90. [CrossRef]
14. Warwick, P.D.; Ruppert, L.F. Carbon and oxygen isotopic composition of coal and carbon dioxide derived from laboratory coal combustion: A preliminary study. *Int. J. Coal Geol.* **2016**, *166*, 128–135. [CrossRef]
15. Hare, S.R.; Mantua, N.J. Empirical evidence for North Pacific regime shifts in 1977 and 1989. *Prog. Ocean.* **2000**, *47*, 103–145. [CrossRef]
16. Chow, G. Tests of equality between sets of coefficients in two linear regressions. *Econometrica* **1960**, *28*, 591–605. [CrossRef]
17. Earth System Science Center. Global Temperature Report, Monthly Average Data. 2022. Available online: https://www.nsstc.uah.edu/data/msu/v6.0/tlt/uahncdc_lt_6.0.txt (accessed on 21 June 2023).
18. Alexander, M.A.; Halimeda, K.; Nye, J.A. Climate variability during warm and cold phases of the Atlantic Multidecadal Oscillation (AMO) 1871–2008. *J. Mar. Syst.* **2014**, *133*, 14–26. [CrossRef]
19. Alheit, J.; Licandro, P.; Coombs, S.; Garcia, A.; Giráldez, A.; Santamaría, M.T.G.; Slotte, A.; Tsikliras, A.C. Atlantic Multi-decadal Oscillation (AMO) modulates dynamics of small pelagic fishes and ecosystem regime shifts in the eastern North and Central Atlantic. *J. Mar. Syst.* **2014**, *131*, 21–35. [CrossRef]
20. Rayner N., A.; Parker, D.E.; Horton, E.B.; Folland, C.K.; Alexander, L.V.; Rowell, D.P.; Kent, E.C.; Kaplan, A. Global analyses of sea surface temperature, sea ice, and night marine air temperature since the late nineteenth century. *J. Geophys. Res.* **2003**, *108*, 4407. [CrossRef]
21. Holzer, M.; DeVries, T. Source-labeled anthropogenic carbon reveals a large shift of preindustrial carbon from the ocean to the atmosphere. *Glob. Biogeochem. Cycles* **2022**, *36*, e2022GB007405. [CrossRef]

atmosphere

Article

Lake Surface Temperature Retrieval Study Based on Landsat 8 Satellite Imagery—A Case Study of Poyang Lake

Xudong Kong [1], Yajun Li [1,*], Lingli Wang [2] and Huijie Liu [2,3]

1 School of Marine Sciences, China University of Geosciences, Beijing 100083, China; 2111220011@email.cugb.edu.cn
2 Beijing SatImage Information Technology Co., Ltd., Beijing 100040, China; wanglingli@satimage.cn (L.W.); liuhj@satimage.cn (H.L.)
3 Land Satellite Remote Sensing Application Center (LASAC), Ministry of Natural Resource, Beijing 100048, China
* Correspondence: liyajun_daliu@cugb.edu.cn

Abstract: Poyang Lake is the largest freshwater lake in China and forms an essential component of the hydrological, nutrient, and carbon cycles, providing various ecosystem services to the local environment. Since changes in Poyang Lake's water temperature can significantly affect the surrounding environment and social development, continuous monitoring of lake temperature changes is required. Traditional water monitoring methods are resource intensive and cannot simultaneously conduct extensive water monitoring. Remote sensing of temperature inversion has the advantages of all-weather, efficient, and large-scale real-time monitoring. Six Landsat 8 images from August to October in 2020 and 2021 were utilized to extract lake surface temperature (LST), and the variations in LST over the two years were analyzed to determine the impact of global climate anomalies on inland lakes. The results indicate that the LST in August and October 2021 was significantly higher than that in the same periods of the previous year, and the temperature difference in October reached 8 °C. In contrast to the overall normal distribution pattern of the water temperature in 2020, 2021 exhibited a relatively concentrated, unimodal distribution pattern. A trend analysis of the driving factors suggests that the LST of Poyang Lake is influenced by the global climate, and the artificial heat sources around the lake clearly alter the distribution characteristics of the LST simultaneously.

Keywords: Landsat 8; water temperature; radiation transfer equation; global climate anomalies

1. Introduction

Water temperature is a crucial factor influencing the ecological environment of rivers, lakes, oceans, and other water bodies. Affected by frequent extreme weather worldwide, water temperature anomalies occasionally occur [1–3]. Poyang Lake, a vital component of the Yangtze River Basin ecosystem, plays a significant role in ecological preservation, climate regulation, and water resource utilization [4]. Therefore, dynamic monitoring of the water temperature in Poyang Lake is of great significance. In recent years, the Poyang Lake area has experienced ecological problems caused by abnormal changes in water temperature, especially the extreme drought in the summer of 2022, resulting in the death of many aquatic organisms and serious destruction of the ecological environment [5]. Traditional in situ temperature measurements are real but sparse with low efficiency, which makes it difficult to analyze and predict the features of water temperature in high-resolution and wide areas. Compared to traditional measurement methods, remote sensing technology can rapidly and accurately extract large-scale water information, offering a long observation time and high accuracy. In recent decades, thermal infrared remote sensing technology has been gradually applied to temperature monitoring in large-scale waters, such as lakes and oceans [6]. However, the unsatisfactory performance of the sensor hardware and satellite orbit affects the temporal and spatial resolutions of remote sensing technologies.

As one of the most widely used remote sensing data sources for long-term water temperature inversion, studies based on Landsat thermal infrared data are relatively mature [7].

Multitemporal TM thermal infrared data have been used to determine the water temperature around the Daya Bay Nuclear Power Station [8]. A numerical forecasting model of the vertical profile of the Yellow Sea water temperature based on remote sensing data [9] has proven that remote sensing technology can effectively monitor temperature problems. Common algorithms for inverting the water surface temperature include the radiative transfer, single-window, and split-window algorithms [10]. The radiative transfer algorithm considers the influence of various atmospheres, correcting and eliminating atmospheric effects by calculating the atmospheric radiation and atmospheric transmittance. This algorithm has a high application accuracy for temperature anomaly monitoring and large-area water temperature inversion. The accuracy of inversion results can be improved significantly when combined with edge effect elimination [11–14]. However, complex calculations and the difficult acquisition of real-time atmospheric vertical contour data affect the accuracy of surface temperature inversion results. Unlike the complex radiative transfer algorithm, the single-window algorithm solely uses thermal infrared data to invert the surface temperature, which can exhibit a relatively high correlation with the in situ measured water temperature to a certain extent [15,16]. Compared to the single-window algorithm, introducing the split-window algorithm further reduces the required parameters for temperature inversion. Moreover, it can achieve high-precision inversion results without requiring precise atmospheric profile data and has gradually become the most widely used algorithm for water surface temperature inversion [17]. In recent years, continuous on-site and remote sensing research has been conducted on the water bodies of Poyang Lake. Comprehensive pollution index evaluation methods and principal component analysis allow for an accurate and objective assessment of the spatiotemporal evolution of water quality [18,19]. Using Landsat TM images through logical operations or water indices can effectively extract water bodies from Poyang Lake [20,21]. Additionally, GF-3 images combined with the KI algorithm to model SAR images and reconstruct the cost function, which can minimize the misclassification rate when the overall classification cost is the lowest, can provide good support for the dynamic monitoring of Poyang Lake [22].

While research on Poyang Lake has been ongoing for many years and various systematic research methods have been developed [23–28], most studies still rely on in situ measured temperature data, which are relatively time-consuming and labor-intensive, making it challenging to acquire water temperature information over a large area simultaneously. Applying remote sensing technology not only efficiently supplements in situ measured data with regional water temperature inversion data, but also saves human and material resources, generating reliable surface water temperature data for the Poyang Lake region. These data extract regional water temperature characteristics and support ecological conservation, environmental management, resource planning, and local production. The purpose of this study was to obtain the water temperature in the study area using data from the Landsat 8 remote sensing satellite, and to explore the effects of global climate anomalies and human factors on inland lakes and their internal relationships by combining climate and human activity data.

2. Research Area and Data Source

2.1. Research Area

During the wet season, Poyang Lake is the largest freshwater lake and becomes the second largest lake in China surpassed only by Qinghai Lake, which is the largest saltwater lake in the country. Poyang Lake is located in the northern part of Jiangxi Province and is a major tributary of the middle and lower reaches of the Yangtze River, serving as a crucial lake with throughflow, discharge, and seasonal importance in the Yangtze River Basin (Figure 1). Poyang Lake is primarily fed by the Gan, Xiu, Xin, Rao, and Fu rivers. From south to north, it converges with the Yangtze River near Shizhong Mountain in Hukou

County, Jiujiang City. At normal water levels (14–15 m), the lake's surface area is 3150 km^2. During high water levels (20 m), it exceeds 4125 km^2, while at low water levels (12 m), it is reduced to only 500 km^2. Poyang Lake plays a crucial role in regulating the water level of the Yangtze River, conserving water resources, improving the local climate, and maintaining the ecological balance in surrounding areas.

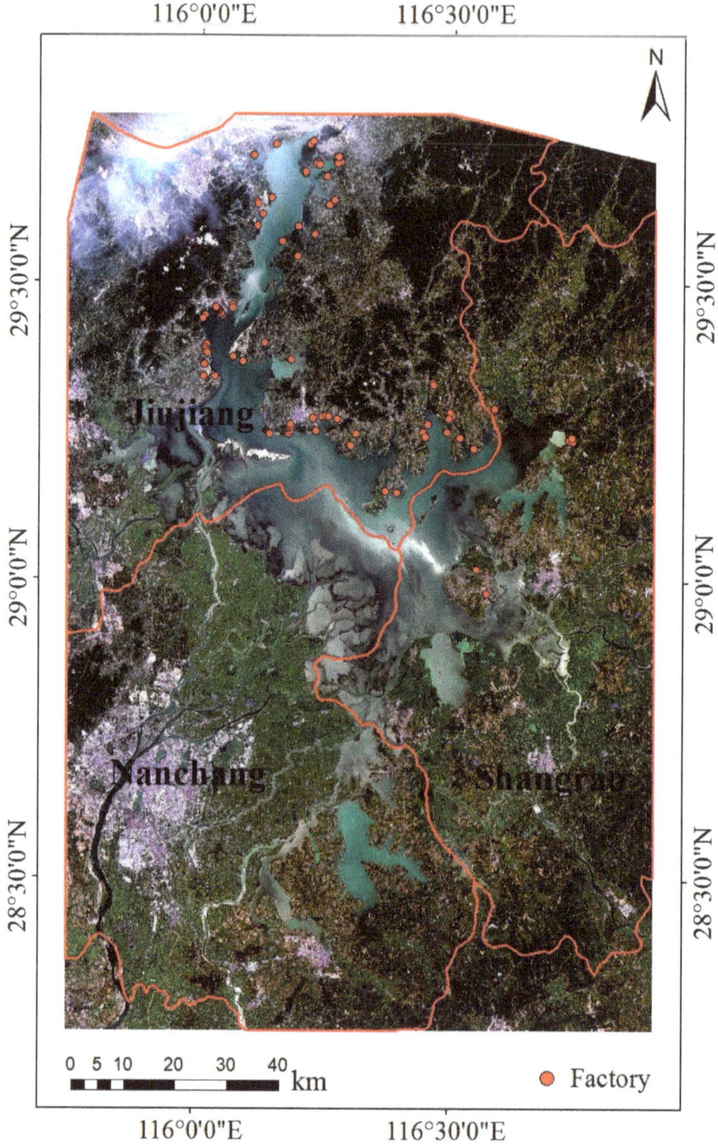

Figure 1. Location of the research area.

2.2. Data Source

The in situ temperature was obtained from water temperature data measured within the jurisdictions of Jiujiang City, Nanchang City, and Shangrao City in the waters of Poyang Lake (https://data.tpdc.ac.cn/en/data/38ea426e-f99c-40c6-87a0-2354419eb2df)

(accessed on 13 September 2023). The satellite remote sensing data were obtained from the Geospatial Data Cloud (https://www.gscloud.cn/) (accessed on 13 September 2023), the United States Geological Survey (https://earthexplorer.usgs.gov) (accessed on 13 September 2023), and the Level-1 and Atmosphere Archive & Distribution System Distributed Active Archive Center (https://ladsweb.modaps.eosdis.nasa.gov) (accessed on 13 September 2023). The temperature, irradiance, wind speed, and precipitation data used for subsequent regional analyses were sourced from the National Centers for Environmental Information (NCEI), a division of the National Oceanic and Atmospheric Administration (NOAA) in the United States (https://www.ncei.noaa.gov/data/global-summary-of-the-day/archive/) (accessed on 13 September 2023).

3. Methods

3.1. Radiative Transfer Equation

In this study, based on the Landsat 8 satellite data, the OLI_TIRS data were preprocessed through radiometric calibration and atmospheric correction. Subsequently, the radiative transfer algorithm was employed to calculate the surface emissivity, leading to the temperature retrieval in the research area (Figure 2).

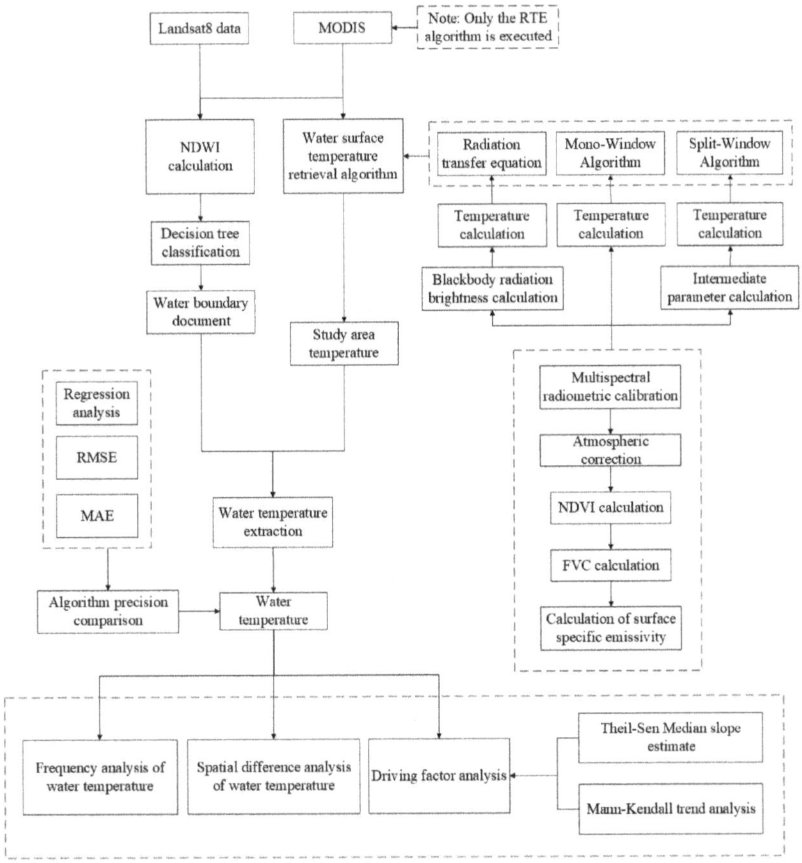

Figure 2. Technical method diagram.

The principle was first to estimate the influence of the atmosphere on the surface thermal radiation, and then remove the atmospheric influence from the total thermal radiation obtained by the sensor, and acquire the surface thermal radiation intensity that could be converted into the surface temperature.

The radiative transfer algorithm first performed radiometric calibration [29] and atmospheric correction [30–32] on the raw data. The vegetation cover was calculated using the normalized difference vegetation index (NDVI) [33] with the following formula:

$$FVC = (NDVI - NDVIs)/(NDVIv - NDVIs) \quad (1)$$

In Equation (1), FVC represents the vegetation coverage, NDVI represents the normalized difference vegetation index, NDVIs is the NDVI value without vegetation cover, and NDVIv is the NDVI value with full vegetation cover [34].

Next, the surface emissivity was calculated using the NDVI threshold method [35] with the following formula:

$$\varepsilon_{surface} = 0.004FVC + 0.0986 \quad (2)$$

In Equation (2), $\varepsilon_{surface}$ represents the surface emissivity, and FVC represents the vegetation coverage.

Following this, the blackbody radiance in the thermal infrared band was computed based on the surface emissivity:

$$B(T_s) = [L_\lambda - L\uparrow - \tau(1 - \varepsilon)\, L\downarrow]/\tau\varepsilon \quad (3)$$

In Equation (3), $B(T_s)$ represents the blackbody radiance; ε is the surface emissivity; T_s is the actual surface temperature; τ is the atmospheric transmittance in the thermal infrared band, $\tau = 0.94$; $L\uparrow$ is the atmospheric upwelling radiance, $L\uparrow = 0.37$; $L\downarrow$ is the atmospheric downwelling radiance, $L\downarrow = 0.66$; and L_λ is the thermal infrared radiance received by the satellite sensor [36,37].

Finally, the surface temperature was inverted using the Planck function with the following formula:

$$Ts = K_2/\ln[K_1/B(Ts) + 1] \quad (4)$$

In Equation (4), T_s is the actual surface temperature; $B(T_s)$ represents the blackbody radiance; and the values of K1 and K2 are determined based on the corresponding sensor, K1 = 774.89, K2 = 1321.08.

3.2. Mono-Window Algorithm

The mono-window algorithm is a retrieval method suitable for data with only one thermal infrared band. In this study, the 10th band in Landsat 8 was selected to use the mono-window algorithm [38] with the following formula:

$$T_S = (a(1 - M - N) + [b(1 - M - N) + M + N]T_{10} - DT_a)/M \quad (5)$$

In Equation (5), T_S represents the surface temperature of the lake, a = −62.73566, b = 0.43404, T_{10} is the radiation brightness temperature of band 10, T_a is the average atmospheric temperature (K), and M and N are the intermediate variables.

There is a linear relationship between the atmospheric average temperature T_a and the near-surface temperature T_0.

$$T_a = 0.92621T_0 + 16.0110 \quad (6)$$

In Equation (6), T_a is the average atmospheric temperature (K), and T_0 is the near-surface temperature.

The two parameters M and N can be calculated using the following method:

$$M = \varepsilon\tau \quad (7)$$

$$N = (1 - \tau)[1 + (1 - \varepsilon)\tau] \quad (8)$$

In Equations (7) and (8), ε is the surface emissivity and τ is the atmospheric transmittance in the thermal infrared band, where $\tau = 0.94$.

3.3. Split-Window Algorithm

The surface temperature inversion equation of the split-window algorithm [39] applicable to Landsat 8 remote sensing image data is shown as follows:

$$T_R = T_{10}A_0 - T_{11}A_1 + A_2 \tag{9}$$

In Equation (9), T_R represents the surface temperature of the lake, T_{10} and T_{11} are the radiation brightness temperature of band 10 and band 11, and A_0, A_1 and A_2 are the parameters.

The formula for A_0, A_1, and A_2 is as follows:

$$A_0 = 1 + ([N_{10} + b_{10}N_{11}(1 - M_{10} - N_{10})]/(M_{10}N_{11} - M_{11}N_{10})) \tag{10}$$

$$A_1 = [N_{10} + b_{11}N_{10}(1 - M_{11} - N_{11})]/(M_{10}N_{11} - M_{11}N_{10}) \tag{11}$$

$$A_2 = [a_{10}N_{11}(1 - M_{10} - N_{10}) - a_{11}N_{10}(1 - M_{11} - N_{11})]/(M_{10}N_{11} - M_{11}N_{10}) \tag{12}$$

In Equations (10)–(12), a_{10}, b_{10}, a_{11}, and b_{11} are constants, since the temperature during the study period was between 10 and 40 °C. The constants were $a_{10} = -64.6081$, $b_{10} = 0.4399$, $a_{11} = -69.0215$, and $b_{11} = 0.4756$, here used to calculate bands 10 and 11, corresponding to M and N.

$$M_i = \varepsilon_i\ \tau_i \tag{13}$$

$$N_i = [1 - \tau_i]\ [1 + (1 - \varepsilon_i)\ \tau_i] \tag{14}$$

In Equations (13) and (14), i represents two bands, τ_i is the atmospheric transmittance in the thermal infrared band, $\tau_{10} = 0.94$, $\tau_{11} = 0.92$, and ε_i is the surface emissivity formula, as follows:

$$\varepsilon_i = P_V\ R_V\ \varepsilon_{iv} + (1 - P_V)\ R_S\ \varepsilon_{is} + d_\varepsilon \tag{15}$$

In Equation (15), i represents two bands, ε_i is the surface emissivity, ε_{iv} and ε_{is} are the surface emissivity of vegetation or bare land in the i band, $\varepsilon_{10v} = 0.98672$, $\varepsilon_{10s} = 0.96767$, $\varepsilon_{11v} = 0.98990$, $\varepsilon_{11s} = 0.97790$, P_V is the vegetation cover, R_V and R_S represent the proportion of temperature of vegetation and bare land, and d_ε is a thermal radiation correction term, resulting from the interaction of thermal radiation between vegetation and bare ground, the formula for which is as follows:

$$R_V = 0.0585P_V + 0.9332 \tag{16}$$

$$R_S = 0.1068P_V + 0.9902 \tag{17}$$

$$d_\varepsilon = 0.003796 \min\ (P_V, 1 - P_V) \tag{18}$$

In Equations (16)–(18), R_V and R_S represent the proportion of temperature of vegetation and bare land, P_V is the vegetation cover, and d_ε is a thermal radiation correction term, resulting from the interaction of thermal radiation between vegetation and bare ground.

3.4. Normalized Water Index Method

The normalized water index method involves classifying water bodies and conducting a comprehensive assessment by aggregating the evaluation indicators of various water bodies into an overall index. It primarily uses specific bands in remote sensing imagery for normalized difference processing to highlight water body information in the image.

$$NDWI = (Green - NIR)/(Green + NIR) \tag{19}$$

In Equation (19), green represents the green band corresponding to the second band in the TM images and the third band in the OLI images, whereas NIR represents the near-infrared band corresponding to the fourth band in the TM images and the fifth band in the OLI images [40].

3.5. Decision Tree

A simple binary decision tree was constructed using a decision tree to classify the normalized water body index images. Appropriate thresholds were initially set, and the image data were imported for classification to obtain the water body boundaries [41,42].

3.6. Theil–Sen Median Slope Estimation and Mann–Kendall Trend Analysis

The Theil–Sen Median method is a robust non-parametric statistical trend calculation method, which is suitable for the trend analysis of long time series data. Its calculation formula is as follows:

$$\beta = \text{Median}\left(\frac{x_j - x_i}{j-i}\right) \forall j > i \tag{20}$$

In Equation (20), Median is the formula for taking the mid-value, and i and j are the numbers of data. If $\beta > 0$, it is an increasing trend, and if $\beta < 0$, it is a decreasing trend.

The Mann–Kendall test is a non-parametric time series trend test method, which does not require the measured values to follow the positive distribution, is not affected by missing values and outliers, and is suitable for the trend significance testing of long time series data, the formula for which is as follows:

$$S = \sum_{i=1}^{n-1} \sum_{j=i+1}^{n} \text{sgn}(x_j - x_i) \tag{21}$$

$$\text{sgn}(x_j - x_i) = \begin{cases} +1 \ x_j - x_i > 0 \\ 0 \ x_j - x_i = 0 \\ -1 \ x_j - x_i < 0 \end{cases} \tag{22}$$

In Equations (21) and (22), sgn is a symbolic function, S is the test statistic, and i and j are the numbers of data.

Z is the test statistic, for which the formula is as follows:

$$Z = \begin{cases} \frac{S}{\sqrt{\text{Var}(s)}} \ (S > 0) \\ 0 \ (S = 0) \\ \frac{S+1}{\sqrt{\text{Var}(s)}} \ (S < 0) \end{cases} \tag{23}$$

$$\text{Var}(S) = \frac{n(n-1)(2n+5)}{18} \tag{24}$$

In Equations (23) and (24), n is the number of data in the sequence, S is the test statistic, and Var is the intermediate variable.

4. Results

4.1. Comparison of Methods

The radiative transfer equation, mono-window, and split-window algorithms were used to retrieve the lake surface temperature of the study area in January and June 2021 based on the Landsat 8 data, and the radiative transfer equation algorithm was used to invert the lake surface temperature of January and June 2021 in the study area based on the MODIS data. Then, the four results were compared.

The surface temperature of Poyang Lake obtained by the inversion of the three algorithms and the surface temperature based on the MODIS data were analyzed by means of linear regression. The results of the regression analysis are shown in Figure 3. The negative correlation coefficients between the inversion values and the actual values of the four results were RTE (0.983), MW (0.982), SW (0.978), and MODIS (0.971). In Table 1, the RMSE (1.75 K) and MAE (2.16 K) of the RTE algorithm inversion results were the smallest, while the RMSE (4.25 K) and MAE (4.57 K) of the SW algorithm inversion results are the largest.

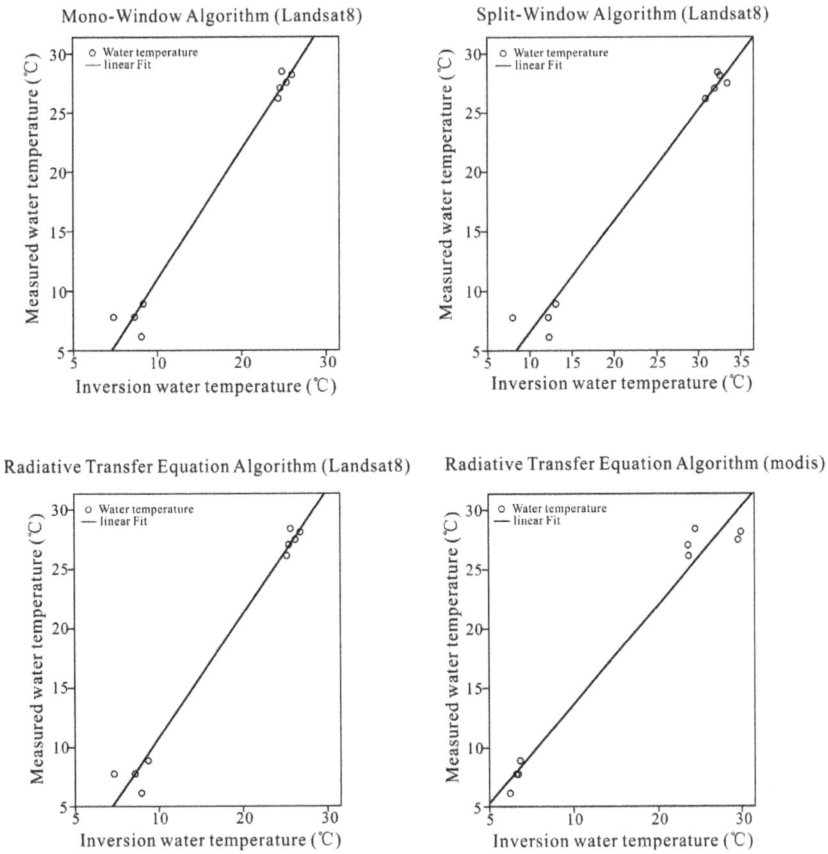

Figure 3. Water temperature fitting results.

Table 1. Method precision comparison.

Method (Data Source)	RMSE (K)	MAE (K)
RTE (Landsat 8)	1.73	1.49
MW (Landsat 8)	2.29	2.04
SW (Landsat 8)	4.40	4.33
RTE (MODIS)	3.70	3.52

Note: RMSE is the root mean square error, MAE is the mean absolute error, RTE is the radiative transfer equation algorithm, MW is the mono-window algorithm, SW is the split-window algorithm.

According to the comprehensive statistical parameters, the inversion result of the RTE algorithm was the closest to the measured value and had the highest inversion accuracy, while the inversion result of the SW algorithm deviated farthest from the measured value and had the lowest inversion accuracy.

A significant correlation between the inverted lake surface temperature (LST) and the measured water temperatures was observed (Figure 3). The temperature inversion results in this study effectively represent the surface temperature variations in the water body. The LST inversion results for Poyang Lake show that the LST in 2021 was generally higher than that in the same period in 2020, with only a slight decrease in temperature observed in September (Figure 4).

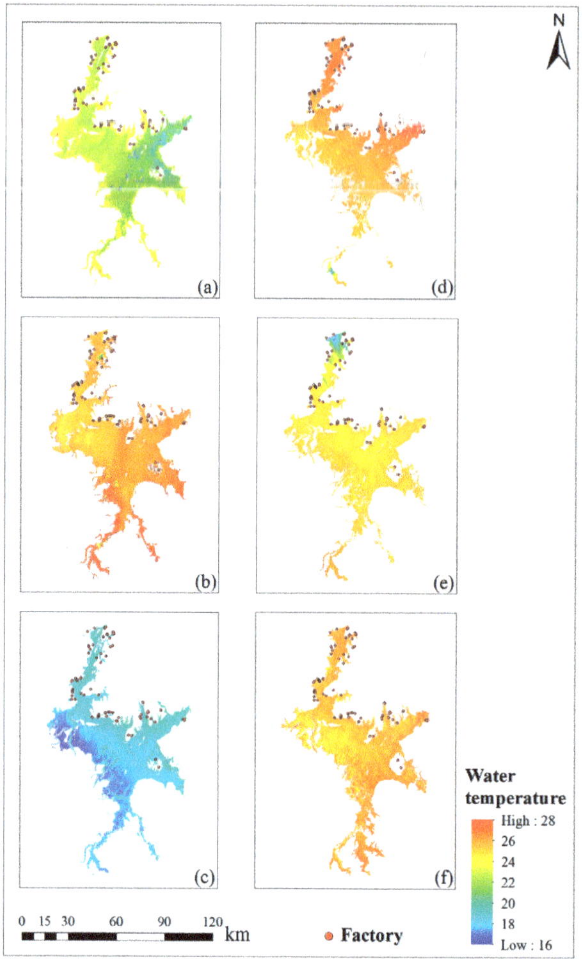

Figure 4. Water temperature data. (**a**) August 2020 (**b**) September 2020 (**c**) October 2020 (**d**) August 2021 (**e**) September 2021 (**f**) October 2021.

4.2. Frequency Distribution

The composition and characteristics of the LST during the different periods in Poyang Lake were analyzed by conducting a frequency analysis of the existing inverted temperature results (Figure 5). The LST in August 2020 and August 2021 were concentrated between 20 and 24 °C and 25 and 26 °C, respectively, with the temperatures in August 2021 being significantly higher than those during the same period in the previous year (Figure 5a,d). In September 2020, the LST was concentrated between 24 and 27 °C, while in September 2021, it was concentrated between 24 and 25 °C, with a small difference between the two years (Figure 5b,e). In October 2020, the LST was concentrated between 17 and 19 °C, whereas in October 2021, it was concentrated in the range of 24–25 °C, indicating a striking difference between the two years (Figure 5c,f). The LST in 2020 exhibited an overall normal distribution pattern, whereas the LST distribution in 2021 was relatively concentrated, showing a unimodal distribution. Unlike the gradual decrease in LST from August to October 2021, the LST in September 2020 was significantly higher than that in August 2020 and September 2021. The imagery acquisition time for September 2020 was at the beginning of the month, which was influenced by local temperature warming, and there was a slight increase in the LST of Poyang Lake (Figure 5).

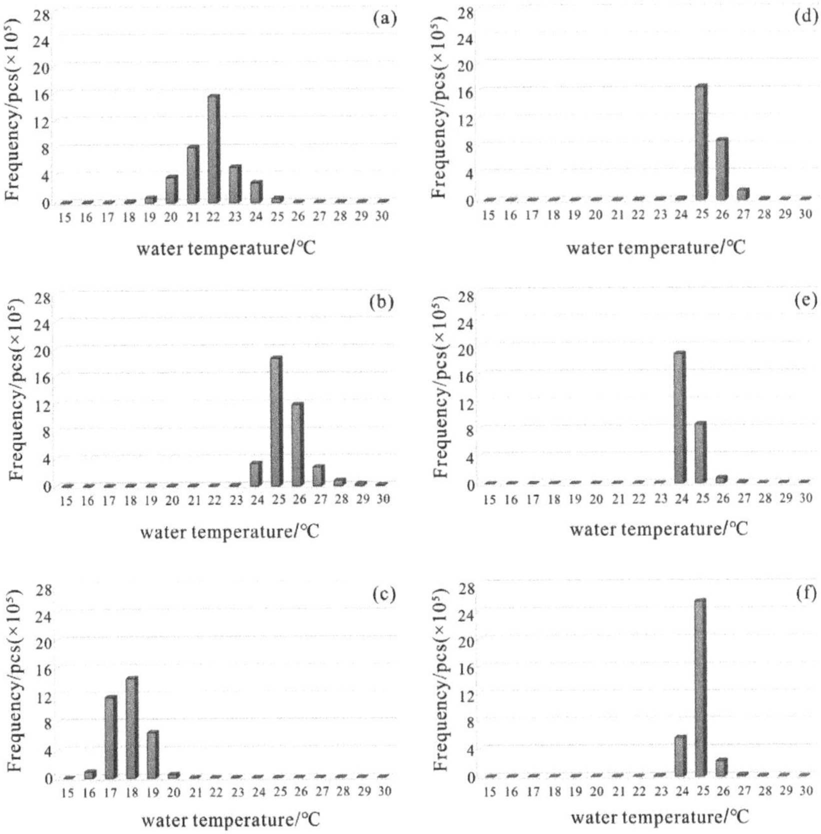

Figure 5. Monthly water temperature frequency. (**a**) August 2020, (**b**) September 2020, (**c**) October 2020, (**d**) August 2021, (**e**) September 2021, (**f**) October 2021.

5. Discussion

5.1. Relationship between Water Temperature and Air Temperature

The average daily air temperatures for August and October 2020–2021 showed a decrease in air temperature from August to September 2020 (Figure 6), which is inconsistent with the inverting trend of the rising LST. This discrepancy is primarily attributed to the fact that the imagery for September 2020 was acquired on 6 September, with a daily average air temperature of 24.3 °C. The overall climate in the lake area was still warm in August, and the temperature had not decreased.

In August and September 2021, the monthly average temperature slightly decreased, with the monthly average temperature being 17.5 °C in October. Compared to the previous month, there was a significant temperature drop of up to 8.4 °C, contrasting with the inverted LST results. October's inverted LST was slightly higher than that of the previous month and significantly higher than the local monthly average temperature, indicating a deviation from the overall temperature trend. An analysis revealed that the imagery for October 2021 was captured on 3 October, with a daily average air temperature of 27.7 °C. Because significant cooling had not yet occurred, there was a noticeable deviation between the inverted LST results and the monthly average temperature for that month. The inverted LST in August 2020 was lower than that of the same period in August 2021, deviating from the local temperature trend, which is also attributed to the influence of the imagery acquisition time.

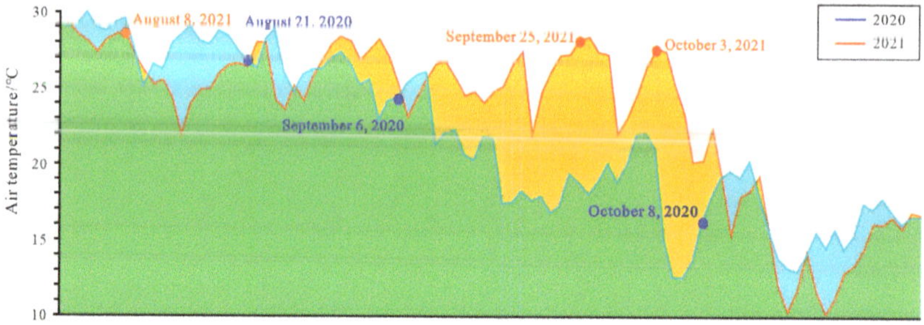

Figure 6. Air temperature trend.

By analyzing the daily average temperatures for 2020 and 2021 (Figure 6), along with the climate anomaly phenomena (Table 2), it was observed that although there were some fluctuations in the trend of temperature change, the overall declining trend corresponded to the occurrence of La Niña. This indicates that global climate anomalies can affect inland temperatures. The frequency analysis of the LST and daily average air temperature for 2020 and 2021 indicated a corresponding relationship between the LST and air temperature. The LST can, to some extent, reflect the local air temperature. However, owing to the large specific heat capacity of water, LST has a certain lag compared to air temperature.

Table 2. Climate anomalies from 2020 to 2021.

Period	Type	Time Length (Months)	Peak Time	Peak Intensity (°C)	Strength Grade
August 2020–March 2021	La Niña	8	November 2020	−1.3	Moderate
September 2021–January 2023	La Niña	17	April 2022	−1.2	Weak

Note: Cited from the Historical Time Statistics Table of the National Climate Center on ENSO (El Niño–Southern Oscillation).

Gifty Attiah et al. applied a retrieval algorithm to the thermal bands of the Landsat archives to generate an LST dataset (North Slave LST dataset) for 535 lakes in the North Slave Region (NSR) of the Northwest Territories (NWT), Canada, for the period of 1984 to 2021. According to the generated North Slave LST, the warm lakes are mainly located around Yellowknife and in the southwest of the NSR.

The authors mentioned that lakes have been treated as homogenous entities in some studies. However, there is spatial variability in the surface temperature of lakes, which depends on several factors, including morphological differences or biological, physical, and human activities occurring on the lake at a given time [43].

Among other things, the authors also raised issues that affect the accuracy of the data, primarily including potential mixed pixels that may not be captured by the algorithm, the presence of "no data" pixels on the lake, and an inconsistent temporal resolution for each lake dataset [44].

5.2. Driving Factor Trend Analysis

Data on the temperature, precipitation, wind speed, and global irradiance were collected from the three cities where Poyang Lake spans to further determine the relationship between lake water temperature and various natural elements. Theil–Sen Median slope estimation and Mann–Kendall trend analysis [45–47] were carried out to calculate the trends in the elements for each month and year. The temperature, precipitation, and wind speed trends were almost consistent across the different cities during the same period, indicating a basic similarity in climatic conditions.

In 2021, the overall irradiance increased (Table 3), whereas the temperature gradually decreased. This phenomenon can be attributed to the transitional study period between summer and autumn. The decreasing precipitation trend was due to the global occurrence of warm and dry climatic conditions, which resulted in less precipitation. The increased irradiance and decreased temperature from 2020 to 2021 are speculated to be related to the La Niña climate anomaly, characterized by cold and dry conditions, which is consistent with the trend analysis results for temperature and precipitation.

Table 3. Trend table of driving factors.

Time \ City / Atem	Jiujiang	Nanchang	Shangrao
2020 (day)	−1	−1	−1
2021 (day)	−1	−1	−1
Time \ City / Irr	Global scale		
2020 (day)	−1		
2021 (day)	1		
Time \ City / Ws	Jiujiang	Nanchang	Shangrao
2020 (day)	−1	−1	−1
2021 (day)	1	1	1
Time \ City / Precip	Jiujiang	Nanchang	Shangrao
2012–2021 (month)	−1	−1	−1

Note: −1 represents no significant decrease, and 1 represents no significant increase; Atem represents air temperature, Irr represents irradiance, Ws represents wind speed, Precip represents precipitation.

Landsat 8 OLI satellite data with a resolution of 30 m were used in this study. Limited image resolution may result in the omission of some small water bodies, leading to errors in water body extraction and influencing the research results. It is advisable to use high-resolution imagery more frequently in future studies to enhance the accuracy of water body identification.

5.3. Spatial Differences in Water Temperature

The existing inverted LST results were reclassified to visually display the spatial differences in the water temperature of Poyang Lake for better analysis of the regional temperature characteristics (Figure 7). The LST exhibited zonal spatial characteristics unrelated to regional climate variations from August to October 2020 and September 2021 (Table 3). An investigation of local human activities revealed that factories in the Poyang Lake Basin are mainly concentrated along the northeast coast at a high density. In contrast, the southwest coast, because it is submerged in lake water during the flood season, has more muddy areas and relatively inconvenient transportation, with almost no distribution of factories. These results suggest that the continuous input of heat from the factories along the northern coast was the reason for the regional temperature difference on both sides, causing generally higher water temperatures northeast of Poyang Lake than southwest. There was no apparent boundary in water temperature for September and October 2021 due to the strict epidemic control measures in 2021. Coastal factories mostly halted production, and the climatic conditions primarily influenced the water temperature. Therefore, the LST across various locations in Poyang Lake tended to be more uniform.

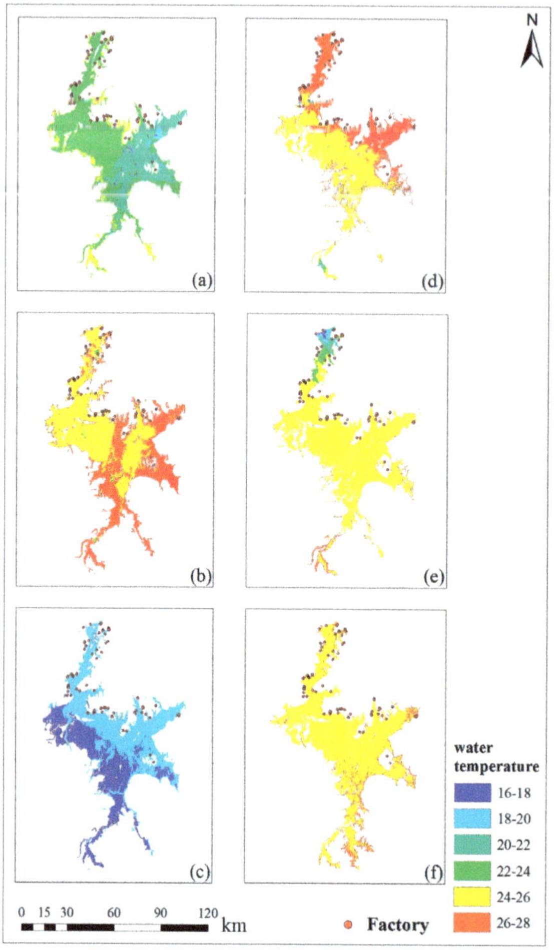

Figure 7. Water temperature reclassification results. (**a**) August 2020 (**b**) September 2020 (**c**) October 2020 (**d**) August 2021 (**e**) September 2021 (**f**) October 2021.

6. Conclusions

The radiative transfer equation, mono-window, and split-window algorithms were used to retrieve the lake surface temperature of the study area based on the Landsat 8 data, and the radiative transfer equation algorithm was used to invert the lake surface temperature of the study area based on the MODIS data. Then, the four results were compared. The radiative transfer equation algorithm based on Landsat 8 has the highest accuracy and reliable results.

This study utilized a radiative transfer equation algorithm to invert the lake surface temperature from Landsat 8 OLI images from six periods, obtaining the LST of Poyang Lake at a large spatial scale for the same periods. By conducting a frequency analysis of the inverted LST, reclassification analysis, temperature correlation analysis, and Mann–Kendall trend analysis on various meteorological data, including irradiance, temperature, precipitation, and wind speed across different cities in the Poyang Lake region, the following conclusions were drawn:

(1) In August 2021, the LST was significantly higher than in the same period in the previous year. The difference in LST was relatively small between September 2020 and

September 2021, while the LST in October 2021 was markedly higher than that in the same period in the previous year, with a temperature difference of 8 °C. Unlike the gradual decrease in LST from August to October 2021, the LST in September 2020 was significantly higher than that in August and the same period in September 2021. Compared with the overall normal distribution pattern of the LST in 2020, the distribution in 2021 showed a relatively concentrated unimodal pattern.

(2) The correlation between LST and air temperature indicates that the surface temperature of inland water bodies is influenced to some extent by regional air temperature. Although there were slight differences in the temperature trends between 2020 and 2021, the overall decreasing trend corresponds to the occurrence of La Niña, suggesting that global climate anomalies impact lake surface temperatures.

(3) The spatial distribution of the LST from August to October 2020 and August 2021 exhibited higher temperatures in the northeast compared to the southwest. The temperatures were higher closer to the shore, and the continuous heat input from the factories along the northern coast is believed to have caused the regional differences in the water temperature between the two shores. Owing to the strict epidemic control measures in September and October 2021, coastal factories were mostly shut down, resulting in the water temperatures being primarily influenced by the regional climate and exhibiting regional temperature homogeneity.

Author Contributions: Methodology, X.K.; Software, L.W.; Resources, H.L.; Data curation, L.W. and H.L., Writing—Original draft, X.K., Writing—Review & editing, Y.L.; Supervision, Y.L.; Project administration, Y.L. All authors have read and agreed to the published version of the manuscript.

Funding: This work was supported by the "National Natural Science Foundation of China" [grant number 41902156], and China Geological Survey [project code: DD20191006].

Institutional Review Board Statement: Not applicable.

Informed Consent Statement: Not applicable.

Data Availability Statement: The data presented in this study are available in this article.

Conflicts of Interest: Lingli Wang and Huijie Liu were employed by the company Beijing SatImage Information Technology Co., Ltd. The remaining authors declare declare that they have no known competing financial interests or personal relationships that could have appeared to influence the work reported in this paper.

References

1. Sun, J. Temporal and Spatial Variation of Lake Water Temperature in the Lower Reaches of the Yangtze River and Its Response to Climate Warming. Master's Thesis, Tianjin University, Tianjin, China, 2021. (In Chinese).
2. Li, Q. Variation characteristics of surface water temperature in Qinghai Lake and its response to climate change. *Yellow River* **2018**, *40*, 25–29. (In Chinese)
3. Wang, X.; Shi, K.; Zhang, Y.; Qin, B. Response of lake surface water temperature to climate change: Past, present and future. *Chin. Sci. Bull.* **2023**, *68*, 3002–3004. (In Chinese) [CrossRef]
4. Li, H.; Xu, Z.; Zhou, Y.; He, X.; He, M. Flood Monitoring Using Sentinel-1 SAR for Agricultural Disaster Assessment in Poyang Lake Region. *Remote Sens.* **2023**, *15*, 5247. [CrossRef]
5. Zhang, Q.; Xue, C.; Xia, J. The impact, causes, and countermeasures of extreme drought in Poyang Lake. *Bull. Chin. Acad. Sci.* **2023**, *38*, 1894–1902. (In Chinese) [CrossRef]
6. CHerrick, C.; Steele, B.G.; Brentrup, J.A.; Cottingham, K.L.; Ducey, M.J.; Lutz, D.A.; Palace, M.W.; Thompson, M.C.; Trout-Haney, J.V.; Weathers, K.C. lakeCoSTR: A tool to facilitate use of Landsat Collection 2 to estimate lake surface water temperatures. *Ecosphere* **2023**, *14*, e4357. [CrossRef]
7. Yan, F.; Wu, L.; Wang, S.; Zhou, Y.; Xu, C.; Wang, L. Review of research on inversion of water surface temperature. *Geo-Inf. Sci.* **2015**, *17*, 969–978. (In Chinese)
8. Wu, C.; Wang, Q.; Wang, W.; Wang, C. Monitoring and Evaluating Therm-water Pollution of Dayawan by TM Images. *Environ. Monit. China* **2006**, *22*, 80–84. (In Chinese) [CrossRef]
9. Xiu, S.; Pang, A.; Shen, H.; Zhang, Q. Remote Sensing Inversion and Numerical Prediction Model for the Temperature Vertical Profile in the Yellow Sea. *Adv. Mar. Sci.* **2009**, *27*, 444–451. (In Chinese)
10. Shi, X.; Xia, J.; Sun, J. Comparison of methods to derive river water temperature using thermal infrared imagery: A case study of the upper Yangtze River catchment. *J. Lake Sci.* **2022**, *34*, 307–319. (In Chinese)

11. Gong, S.; Sun, H.; Wang, S.; Guo, W.; Li, Y. Study on atmospheric transmittance of thermal infrared remote sensing(II): Application of atmospheric transmittance model. *Infrared Laser Eng.* **2015**, *44*, 2013–2020. (In Chinese)
12. Shi, X.; Sun, J.; Shi, L. Derivation of river surface temperature from Landsat thermal infrared data. *J. Hydroelectr. Eng.* **2021**, *40*, 121–130. (In Chinese)
13. Yuan, L. A comparative study of IRS and TM Image Monitoring of Thermal Pollution in Daya Bay. Master's Thesis, CUGB, Beijing, China, 2013. (In Chinese).
14. Qi, J.; Li, X. Research on Retrieval of Surface Temperature Based on Landsat8 Data Atmospheric Correction Method. *Acad. J. Environ. Earth Sci.* **2022**, *4*, 22–27. [CrossRef]
15. Huang, M.; Mao, Z.; Xing, X.; Sun, Z.; Zhao, Z.; Huang, W. A Model for Water Surface Temperature Retrieval from HJ-1B/IRS Data and Its Application. *Remote Sens. Land Resour.* **2011**, *6*, 81–86. (In Chinese)
16. Zhang, J.; Liu, H.; Liang, D. Comparison of two algorithms for water temperature of Taihu Lake based on HJ/IRS data. *J. Shaanxi Meteorol.* **2012**, *5*, 25–28. (In Chinese)
17. Lv, C. Temperature Drainage Monitoring of Hongyanhe Nuclear Power Plant Based on Landsat8 Data Split Window Algorithm. Master's Thesis, Liaoning Technical University, Fuxin, China, 2016. (In Chinese).
18. Liu, W.; Chu, Y.; Tan, Q.; Chen, J.; Yang, L.; Zhou, Q.; Wu, Z.; He, F. Distributional characteristics and evaluation of nitrogen, phosphorus and organic matter in the surface sediments of rural ponds around the western part of Lake Poyang in summer. *J. Lake Sci.* **2023**, *35*, 192–202. (In Chinese)
19. Xie, H.; Hu, M.; Ji, X.; Cao, B.; Jia, S.; Xu, J.; Jin, X. Water Quality Evolution Characteristics and Pollution Factor Analysis in Poyang Lake from 2011 to 2019. *Environ. Sci.* **2022**, *43*, 5585–5597. (In Chinese) [CrossRef]
20. Zhao, H.; Chen, Z.; Feng, C.; Zhang, T.; Zhao, X.; Zhang, Z. Monitoring of Poyang Lake area and its relationship with water level based on Landsat images. *Remote Sens. Nat. Resour.* **2023**, *9*, 1–9. (In Chinese) [CrossRef]
21. Dong, L.; Yan, L.I. The Calculation of Area and Storage of Poyang Lake Based on Remote Sensing Technology. *Remote Sens. Inf.* **2012**, *2*, 57–61.
22. Li, L.; Cui, Q.; Zhang, Z.; Li, X.; Chen, D.; Zha, C.; Wang, X. Dynamic Monitoring and Impact Assessment of Poyang Lake Flood in 2020 Based on GF-3. *Hydrology* **2022**, *42*, 43–47. (In Chinese) [CrossRef]
23. Feng, L.; Hu, C.; Chen, X.; Tian, L.; Chen, L. Human induced turbidity changes in Poyang Lake between 2000 and 2010: Observations from MODIS. *J. Geophys. Res. Ocean.* **2012**, *117*. [CrossRef]
24. Han, X.; Chen, X.; Feng, L. Four decades of winter wetland changes in Poyang Lake based on Landsat observations between 1973 and 2013. *Remote Sens. Environ.* **2015**, *156*, 426–437. [CrossRef]
25. Michishita, R.; Jiang, Z.; Xu, B. Monitoring two decades of urbanization in the Poyang Lake area, China through spectral unmixing. *Remote Sens. Environ.* **2011**, *117*, 3–18. [CrossRef]
26. Ye, X.C.; Zhang, Q.; Liu, J. Impacts of Climate Change and Human Activities on Runoff of Poyang Lake Catchment. *J. Glaciol. Geocryol.* **2009**, *31*, 835–842.
27. Yu, Z.; Chen, X.; Zhou, B.; Tian, L.; Yuan, X.; Feng, L. Assessment of total suspended sediment concentrations in Poyang Lake using HJ-1A/1B CCD imagery. *Chin. J. Oceanol. Limnol.* **2012**, *30*, 295–304. [CrossRef]
28. Feng, L.; Hu, C.; Chen, X.; Li, R.; Tian, L.; Murch, B. MODIS observations of the bottom topography and its inter-annual variability of Poyang Lake. *Remote Sens. Environ.* **2011**, *115*, 2729–2741. [CrossRef]
29. Hossain, A.K.M.A.; Blanton, R.; Mathias, C.A. Quantitative remote sensing of surface water quality in southeast tennessee using planet dove imagery. In Proceedings of the Joint 69th Annual Southeastern/55th Annual Northeastern GSA Section Meeting-2020, Reston, VA, USA, 20–22 March 2020.
30. Li, H. Study on Applicability Evaluation of Atmospheric Correction Method in Poyang Lake. Master's Thesis, Jiangxi University of Science and Technology, Ganzhou, China, 2022. (In Chinese).
31. Liu, Z.; Zhao, S.; Pei, L.; Liu, S.; Wang, X. Precision Evaluation on Atmospheric Correction Reflectance of GF-5 Satellite Hyperspectral Data. *Remote Sens. Inf.* **2021**, *36*, 93–98. (In Chinese)
32. Bernardo, N.; Watanabe, F.; Rodrigues, T.; Alcântara, E. Atmospheric correction issues for retrieving total suspended matter concentrations in inland waters using OLI/Landsat-8 image. *Adv. Space Res.* **2017**, *59*, 2335–2348. [CrossRef]
33. Carlson, T.N.; Ripley, D.A. On the relation between NDVI, fractional vegetation cover, and leaf area index. *Remote Sens. Environ.* **1997**, *62*, 241–252. [CrossRef]
34. Jing, Z.; Jing, L.; Qinhuo, L.; Baodong, X.; Xihan, M.; Yadong, D. Generation of a 16 m/10-day fractional vegetation cover product over China based on Chinese GaoFen-1 observations: Method and validation. *Int. J. Digit. Earth* **2023**, *16*, 4229–4246.
35. Wei, X.; Wu, Z.; Wang, R. Research on heat island effect in Nanguan District of Changchun City based on remote sensing. *Jilin Norm. Univ. J.* **2021**, *42*, 134–140. (In Chinese) [CrossRef]
36. Liu, Y.; Liu, L.; Liu, J.; Li, Z. Characteristics and Influence Factors of Summer Surface Urban Heat Island under Arid and Semi-arid Climate. *Sci. Technol. Eng.* **2017**, *17*, 160–165. (In Chinese)
37. Qin, Z.H.; Li, W.J.; Xu, B. The estimation of land surface emissivity for landsat TM6. *Remote Sens. Land Resour.* **2004**, *3*, 28–32.
38. Hou, Y.; Zhang, D. Comparative study of land surface temperature inversion methods based on Landsat 8 remote sensing images. *Chin. Agric. Sci. Bull.* **2019**, *35*, 142–147.
39. Lu, X.; Xu, C.; Guo, J. Comparative analysis of land surface temperature inversion methods based on Landsat 8 data. *Arid Environ. Monit.* **2023**, *37*, 137–144.

40. Chen, D.; Jackson, T.J.; Li, F.; Cosh, M.H.; Walthall, C. Estimation of vegetation water content for corn and soybeans with a normalized difference water index (NDWI) using Landsat Thematic Mapper data. In Proceedings of the Geoscience and Remote Sensing Symposium, Toulouse, France, 21–25 July 2003. IGARSS'03 Proceedings.
41. Feyisa, G.L.; Meilby, H.; Fensholt, R.; Proud, S.R. Automated Water Extraction Index: A new technique for surface water mapping using Landsat imagery. *Remote Sens. Environ.* **2014**, *140*, 23–35. [CrossRef]
42. Shuangtong, L.; Mingxiao, W.; Shuwen, Y.; Mingze, Y.; Lihua, Y.; Geomatics, F.O.; University, L.J. Research on Water Body Extraction Method Based on GF-2 High Resolution Remote Sensing Image. *GNSS World China* **2018**, *43*, 37–43.
43. Crosman, E.T.; Horel, J.D. MODIS-derived surface temperature of the Great Salt Lake. *Remote Sens. Environ.* **2008**, *113*, 73–81. [CrossRef]
44. Gifty, A.; Homa, K.P.; Andrea, S.K. Lake surface temperature retrieved from Landsat satellite series (1984 to 2021) for the North Slave Region. *Earth Syst. Sci. Data* **2023**, *15*, 1329–1355.
45. Peng, Z. Characteristics and Driving Factors of Surface Water Temperature Variation in Major Large Lakes in China. Master's Thesis, Yunnan Normal University, Kunming, China, 2022. (In Chinese)
46. Mesias, J.M.; Bisagni, J.J.; Brunner, A.-M.E.G. A high-resolution satellite-derived sea surface temperature climatology for the western North Atlantic Ocean. *Cont. Shelf Res.* **2006**, *27*, 191–207. [CrossRef]
47. Thomas, A.; Byrne, D.; Weatherbee, R. Coastal sea surface temperature variability from Landsat infrared data. *Remote Sens. Environ.* **2002**, *81*, 262–272. [CrossRef]

atmosphere

MDPI

Article

Determination of Transport Pathways and Mutual Exchanges of Atmospheric Moisture between Source Regions of Yangtze and Yellow River Basins

Beiming Kang [1,2,3,4], Jiahua Wei [1,2,3,4,5,*], Olusola O. Ayantobo [6] and Haijiao Yang [1,2,3,4]

1. School of Civil Engineering and Water Resources, Qinghai University, Xining 810016, China; 2014990028@qhu.edu.cn (B.K.); 2014990031@qhu.edu.cn (H.Y.)
2. State Key Laboratory of Plateau Ecology and Agriculture, Qinghai University, Xining 810016, China
3. Laboratory of Ecological Protection and High-Quality Development in the Upper Yellow River, Qinghai University, Xining 810016, China
4. Key Laboratory of Water Ecology Remediation and Protection at Headwater Regions of Big Rivers, Ministry of Water Resources, Xining 810016, China
5. State Key Laboratory of Hydroscience and Engineering, Tsinghua University, Beijing 100084, China
6. College of Civil Engineering and Architecture, Guangxi University, Nanning 530004, China; skollar24k@yahoo.com

* Correspondence: weijiahua@tsinghua.edu.cn

Abstract: Knowledge of the quantitative importance of the moisture transport pathways and mutual moisture exchange of the source regions of the Yangtze (SYZR) and Yellow (SYR) rivers' basins, the adjacent origins of China's two longest rivers, can provide insights into the regional atmospheric branch of the hydrological cycle over the source regions. The method with the Hybrid Single-Particle Lagrangian Integrated Trajectory (HYSPLIT) model and a Lagrangian moisture source diagnostic to identify the major moisture transport pathways quantifies their importance to two types of daily precipitation events—daily precipitation more than 10 mm (PM) events and daily precipitation less than 10 mm (PL) events—for the two rivers' regions during the summer (June–August, 1986–2015) and finds the characteristics of mutual moisture exchange. The results indicated that both the Bay of Bengal group pathway and the northwest China group pathway play significant roles in PM and PL events over the SYZR, contributing 41.87% and 39.12% to PM events and 41.33% and 33.16% to PL events, respectively. The SYR has five main moisture path groups; the Bay of Bengal group pathway, the northwest China group pathway, and the southeast China group pathway play significant roles in PM and PL events over the SYR, contributing 32.34%, 23.28%, and 34.36% to PM events and 34.84%, 36.18%, and 19.83% to PL events, respectively. The volume of moisture passing from the SYZR to the SYR is approximately 60 times that of the reverse, constituting about 6.9% of the total moisture released in SYR precipitation. It is worth noting that the moisture release was concentrated in the nearer west group pathway, and the main moisture uptake locations were beyond the source region of the two rivers (remote sources) in the PM events. The aggregate moisture release high-frequency moisture transport path groups are found in the southeastern parts of Zhiduo County and the southeast of Zaduo County.

Keywords: moisture pathways; moisture exchange; Lagrangian trajectory model; source regions of the Yangtze and Yellow River

1. Introduction

The Yangtze and Yellow Rivers are China's two longest rivers, with both originating from the eastern part of the Tibetan Plateau (TP) and their source regions lying adjacent to each other [1,2]. These source regions constitute vital ecological barriers in Asia, playing a pivotal role in the intricate dynamics of the global water cycle, hydrological equilibrium, and the provisioning of water resources. Approximately 40% of the Yellow River's total runoff and 1.3% of the Yangtze River's total runoff originate from these regions [3–5]. With the warming and humid climate trends over the TP, these source regions are increasingly contributing to river runoff. The average annual precipitation in the SYR is about 502.72 mm, while the SYZR, being colder and drier, receives an average of 343.89 mm of precipitation annually, indicating a significant disparity between the two source regions [6]. As the atmospheric moisture transport process is an important part of the hydrological cycle, the knowledge of where moisture comes from (moisture sources), how it is transported (moisture transport pathways), and how it is exchanged (mutual moisture exchange) is of great value to provide a full picture of the hydrological cycle in the SYZR and the SYR, and this issue deserves a systematic and in-depth study.

A considerable amount of research has been conducted on the atmospheric moisture transport characteristics of precipitation in both the two river source regions and the TP [7]. Previous studies predominantly employed the Eulerian method; however, in recent years, there has been an increasing trend towards utilizing the Lagrangian approach [8,9]. Research primarily focuses on two aspects: atmospheric moisture transport pathways and the contribution of different atmospheric moisture pathways to regional rainfall [10,11]. During summer and autumn, the TP primarily experiences two moisture pathways originating from the Indian Ocean and the Eurasian continent, while during spring and winter, the mid-latitude westerlies serve as the main moisture source pathways for the TP [12]. Due to variations in study areas, seasons, and rainfall intensity, different atmospheric moisture transport pathways exhibit significant differences in their contribution to regional precipitation. Zhang et al. [13] utilized the HYSPLIT model to study the characteristics of moisture transport during the rainy season in the Three-River Headwaters Region (TRHR), identifying northwest and southern China pathways as the main contributors to extreme and moderate precipitation events in the region. The contribution rates of these pathways to the two types of precipitation events were found to be 18.4% and 25.9% for the northwest pathway and 32.2% and 28.5% for the southern China pathway, respectively. In extreme drought events, the far-western Asia pathway plays a dominant role. Xu et al. [12] observed that during the period from 1979 to 2018, the mid-latitude westerlies' contribution to precipitation in the TP increased, while the contribution from the southern Indian Ocean decreased. Tang et al.'s [14] study on moisture source variations in the TP from 1979 to 2013 found that land-sourced moisture accounted for 69%, while ocean-sourced moisture accounted for approximately 21%. Zhu et al. [15] revealed that the vast arid and semi-arid grassland areas north of the TP were the primary moisture sources for precipitation in the Yellow River source area in July 2015, with a contribution rate as high as 52.9%. However, this contribution rate significantly decreased in July 2012. Research on the characteristics of moisture transport in the source regions of the two rivers, which possess abundant atmospheric water resources [16], is scarce despite the inevitable mutual moisture exchange between neighboring river source regions. Research on the atmospheric moisture transport and exchange characteristics between the two river source regions aims to explore the feasibility of the cross-basin allocation of atmospheric moisture resources through artificial weather modification operations. This provides insights into addressing watershed water scarcity through the development of atmospheric moisture resources.

The details of the dataset and methodology are described in Section 2. The results are shown in Section 3. Discussions and the major conclusions are provided in Section 4.

2. Data and Methods

2.1. Study Area

The study area encompasses the SYZR and the SYR, situated in the eastern part of the TP, ranging from 32°12′ N to 36°15′ N and 90°23′ E to 103°28′ E. The respective regions are 142,714 km^2 and 122,971 km^2, with average elevations of 4757.42 m and 4127.06 m. The annual average temperatures are 0.16 °C and 0.86 °C, while the annual average precipitation is 343.89 mm and 502.72 mm. Approximately 80% of the precipitation occurs during the summer. Topographical maps of the two river source regions are presented in Figure 1.

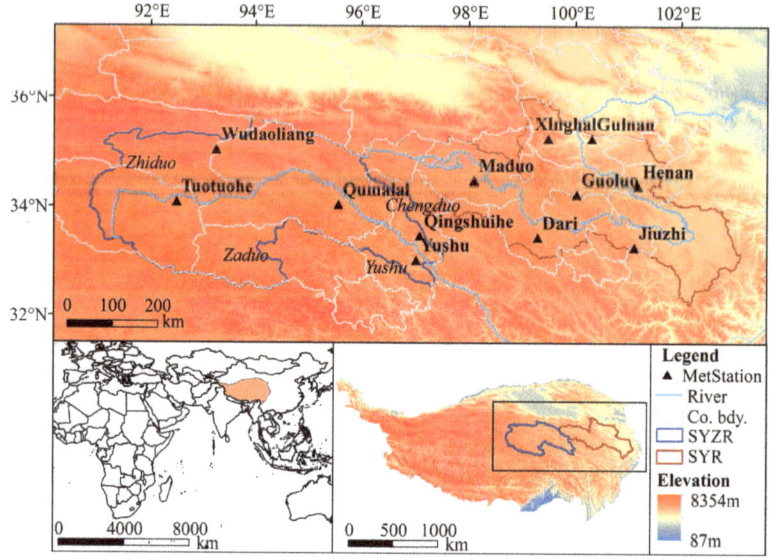

Figure 1. A topographical map of the two river source regions. Co. bdy. stands for the county boundary.

2.2. Definition of Two Types of Daily Precipitation Events

Summer (June–August) daily precipitation data (1986–2015) from 12 meteorological stations located inside the study area (Figure 1) are used to define two types of daily precipitation events. The daily precipitation data are provided by the China National Meteorological Information Center [17]. First, precipitation intensity is classified according to the 24 h precipitation amount categories adopted by the Chinese Meteorological Department [18] (Table 1). The proportion of rainstorms and above precipitation is 0.30%; heavy precipitation accounts for 6.96%, moderate precipitation for 39.38%, and light precipitation for 53.37%.

Table 1. Classification of precipitation types for 24 h.

Grade	Extra Heavy Rainstorm	Heavy Rainstorm	Rainstorm	Heavy Precipitation	Moderate Precipitation	Light Precipitation
24 h Precipitation Amount (mm)	≥250	100–249.9	50–99.9	25–49.9	10–24.9	0.1–9.9
Proportion (%)		0.30		6.96	39.38	53.37

According to the precipitation proportion, precipitation events can be roughly divided into two types: daily precipitation less than 10 mm (PL) events and daily precipitation more than 10 mm (PM) events. Due to the extremely harsh climatic conditions in the study area and the sparse distribution of meteorological stations, gridded daily precipitation data ($0.25^{\circ} \times 0.25^{\circ}$) from the Asian Precipitation-Highly Resolved Observational Data Integration Toward Evaluation of Water Resources [APHRODITE; versions APHRO_MA_V1101 (1960–2007) and APHRO_MA_V1101_EX_R1 (2008–15)] [19] project are used to depict the precipitation conditions of the corresponding precipitation events. Zhang's research [13] indicates satisfactory consistency between the stational data and the APHRODITE data in the spatial and temporal variability of the precipitation over the study area.

2.3. Trajectory Generation Description

The HYSPLIT model, developed by Draxler and Hess [20,21], is used to calculate air parcel trajectories and analyze complex transport phenomena. It has gradually been applied in the study of moisture transport characteristics. The reprocessed NCEP data (HYSPLIT-compatible format) are used to drive the model; the data which are used include surface variables like pressure, temperature at 2 m, and wind components at 10 m, as well as upper-level atmospheric data across various pressure levels from 1000 to 10 hpa, covering heights, temperature, wind direction and speed, vertical wind speed, and relative humidity. Due to the distribution law of atmospheric moisture density, moisture density decreases rapidly with the increase in altitude, and moisture is mainly distributed in the atmosphere below 5 km [22], so a release grid (0.75° × 0.75°) clipped by the regional boundary with 10 vertical levels is defined as the release positions of the trajectories [note a broad vertical range, 0–5 km above ground level (AGL) with an interval of 0.5 km]. The trajectory simulation runs backward in time for 240 h, which is the average residence time of moisture in the atmosphere [23,24]. To cover the entire duration of each precipitation event, the simulation for each trajectory commences at 0000 UTC on the second day following the onset of each precipitation event.

The latitude, longitude, height, pressure, specific humidity, and mixed-layer depth are selected as the output parameters, and the output time interval is 1 h. The total numbers of trajectories generated for the two precipitation types are shown in Table 2. In addition, not all trajectories bring moisture to the starting point of the trajectory, and only those that steadily release moisture in the starting point should be considered valid moisture trajectories. That is, moisture trajectories in the present study are defined as those that have statistically significant ($p < 0.05$) negative trends of moisture content in the target region [13].

Table 2. Description used to identify two types of precipitation events, number of generated trajectories, valid trajectories.

Region	No. of Grids	Event	Mean Precipitation Rate (mm/d)	No. of Events (Average Times/Grid Points)	No. of Traj.	No. of Valid Traj.
Source Region of Yangtze River (SYZR)	26	PM	14.51	48.04	12,560	4203
		PL	4.41	409.19	32,260	13,285
Source Region of Yellow River (SYR)	28	PM	15.25	127.13	29,190	7429
		PL	4.61	669.96	130,970	33,039

PM indicates the daily precipitation is greater than 10 mm events; PL indicates the daily precipitation is less than 10 mm events.

2.4. Trajectory Clustering Method

The HYSPLIT model's built-in clustering method is K-means, and it is incapable of clustering trajectories starting from different points [25]. In this study, trajectory files generated by the HYSPLIT model, with different starting points, are still clustered using the K-means method. The range of cluster numbers is determined by identifying the 'elbow' in the average deviation. Within this range, the silhouette coefficient is calculated, and the number of clusters corresponding to the maximum silhouette coefficient value is selected as the final number of clusters (K) [26], thereby achieving the clustering of trajectories with different starting points. As shown in Figure 2, the average deviation range of K at the 'elbow' is 3–6. When K equals 4, which corresponds to the highest silhouette coefficient value, it is selected as the optimal number of clusters.

Applying the previously mentioned method, the valid trajectories of the SYZR and SYR were clustered, with clustering numbers (N) of 4 and 5, respectively. The average deviation and silhouette coefficients for different clustering numbers are presented in Figure A1 and Table A1 (See Appendix A), respectively.

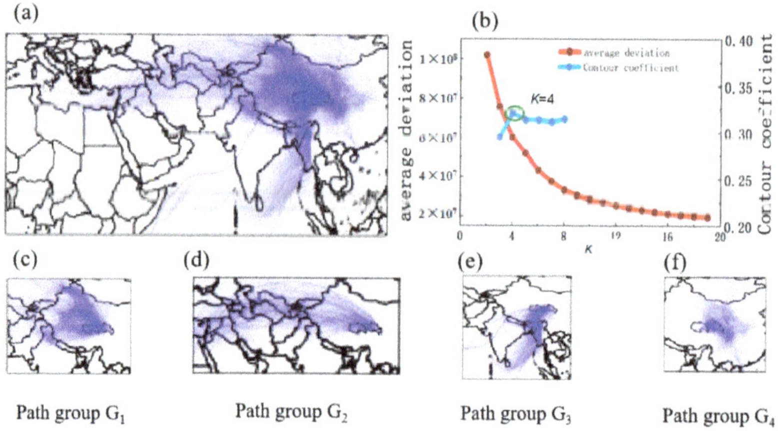

Figure 2. Workflow of trajectory clustering. The blue lines represent trajectories. (**a**) All moisture trajectories; (**b**) number of clusters is determined using average deviation and contour coefficients; K indicates the clustering number. (**c**–**f**) different path groups after clustering moisture trajectories, labeled as path group G_1, path group G_2, path group G_3, and path group G_4.

2.5. Method for Moisture Sources Attribution, Calculating Moisture Path Frequency, and Determining Average Path Humidity

The method for moisture sources' attribution is based on the principles of the S08 and SW14 methods [27,28], which were employed to estimate the moisture contributions from each source region to the precipitation falling in the target area. The sketch of the method for moisture sources' attribution is given in Figure 3.

(1) Identify the moisture uptake and release locations along the moisture trajectory by calculating the moisture changes between neighboring time points. A threshold $e_c = 0.2\ g{\cdot}kg^{-1}\ (6\ h)^{-1}$ is set to select the moisture uptake locations with a moisture uptake amount larger than e_c between the neighboring time points, as performed by Sodemann et al. [28].

(2) Traverse the locations of moisture uptake, loss, and release forward in time. The initial values of the discounted moisture uptake amount Δq are set to $\Delta l_{-n,0}$.

(3) Assume we are at time = $-6(n-1)$, moisture uptake amounts between time = $-6n$ and time = $-6(n-1)$ are $\Delta q_{-6(n-1)}$. The moisture attribution rate by the current uptake location is $f_{-6(n-1)}$. When at time = -12, which is the moisture release time, the portion of the current released moisture amount that comes from the moisture uptake at time = $-6(n-1)$ is $\Delta l_{-6(n-1),-12}$, and the remaining amounts come from the moisture uptake at time = $-6(n-1)$, denoted as $\Delta q_{-6(n-1),-12}$; the current released moisture amount attribution rate that comes from the moisture uptake at time = $-6(n-1)$ is $f_{-6(n-1),-12}$. When at time = 0, which is also a moisture release time, the portion of the current released moisture amount that comes from the moisture uptake at time = $-6(n-1)$ is $\Delta l_{-6(n-1),0}$, and the remaining amounts come from the moisture uptake at time = $-6(n-1)$, denoted as $\Delta q_{-6(n-1),0}$; the current released moisture amount attribution rate that comes from the moisture uptake at time = $-6(n-1)$ is $f_{-6(n-1),0}$.

$$\Delta q_{-6(n-1)} = q_{-6(n-1)} - q_{-6n} \tag{1}$$

$$f_{-6(n-1)} = \frac{\Delta q_{-6(n-1)}}{q_{-6(n-1)}} \tag{2}$$

$$f_{-6(n-1),-18} = \frac{\Delta q_{-6(n-1)}}{q_{-18}} \tag{3}$$

$$\Delta l_{-6(n-1),-12} = \Delta l_{-12} \times f_{-6(n-1),-18} \tag{4}$$

$$\Delta q_{-6(n-1),-12} = \Delta q_{-6(n-1)} - \Delta l_{-6(n-1),-12} \tag{5}$$

$$f_{-6(n-1),-6} = \frac{\Delta q_{-6(n-1),-12}}{q_{-6}} \tag{6}$$

$$\Delta l_{-6(n-1),-0} = \Delta l_{-12} \times f_{-6(n-1),-6} \tag{7}$$

$$\Delta q_{-6(n-1),-0} = \Delta q_{-6(n-1)} - \Delta l_{-6(n-1),-12} - \Delta l_{-6(n-1),-0} \tag{8}$$

(4) For this trajectory, the moisture sources' attribution from remote sources, the SYZR and SYR, is $\Delta l_{-6(n-1),0}$, $\Delta l_{-18,0}$, and $\Delta l_{-6,0}$, respectively. The contributions of different moisture sources can be obtained by accumulating the moisture source attributions of all trajectories for remote sources, the SYZR, and the SYR, denoted as $\Delta l_{\text{remote\_sources}}$, Δl_{SYZR}, and Δl_{SYR}, respectively.

(5) Calculate the total moisture release amount within the target region at time t = 0, denoted as R_{total}. The contribution rates of different moisture sources can be obtained by dividing the moisture source attributions of all trajectories from remote sources, the SYZR, and the SYR by R_{total}.

$$R_{total} = \sum_{i=1}^{L} \Delta q_{0,i} \tag{9}$$

where L is the total number of valid trajectories, and $\Delta q_{0,i}$ is the moisture release amounts between time = −6 and time = 0 of the i-th trajectory.

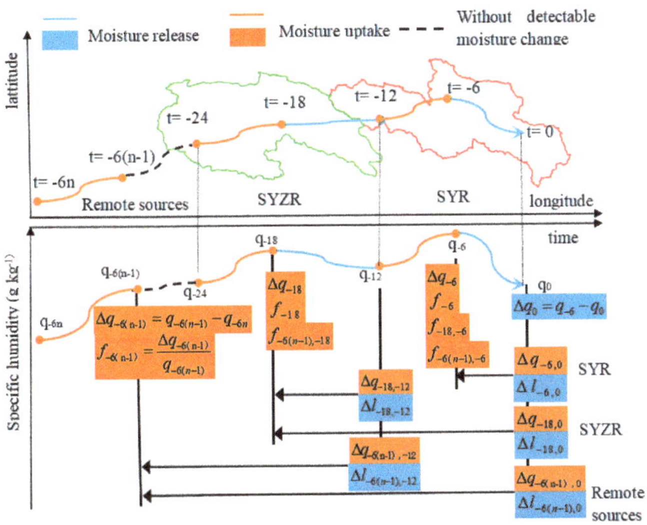

Figure 3. A sketch of the method for moisture sources' attribution. Time points t = −6*n* and −6(*n* − 1) are remote sources; time points t = −24 and −18 are SYZR sources; time points t = −12, −6, and 0 are SYR sources. (**top**) The geographic location of the trajectory. Orange segments and orange lines indicate moisture uptake processes between the neighboring time points, blue segments and blue lines indicate moisture release processes between the neighboring time points, and black segments indicate moisture without detectable change between the neighboring time points. (**bottom**) The specific humidity changes along the trajectory. Orange blocks indicate moisture uptake amounts, and blue blocks indicate moisture release amounts that can be attributed to certain moisture uptake location.

The frequency of moisture trajectories passing over a grid cell (*i,j*) F_{ij} (0.75° × 0.75° grid) is calculated by Equation (10); a higher trajectory frequency indicates that moisture trajectories pass through the grid cell more frequently.

The average specific humidity of moisture trajectories gridded onto each grid cell is calculated by Equation (11). A higher average specific humidity indicates that the trajectories carry abundant moisture.

$$F_{ij} = \frac{M}{L} \tag{10}$$

$$\overline{q_{ij}} = \frac{q_{ij1} + q_{ij2} + \cdots + q_{ijn}}{M} \tag{11}$$

where F_{ij} is the path frequency; L is the total number of valid trajectories; M is the number of paths passing through a grid cell (i, j); $\overline{q_{ij}}$ is the average humidity at a grid cell (i, j); q_{ijn} is the specific humidity of the n-th trajectory passing through the grid cell (i, j).

3. Results

3.1. Major Moisture Transport Path Groups

Figures 4 and 5 show the clustering results of moisture trajectories. In PM events of the SYZR (Figure 4a–d), there are four path groups. Path group1 (SYZR PM1, farther west group) started from a location much farther than West Asia and includes 400 valid trajectories. Path group2 (SYZR PM2, nearer west group) started from West Asia and is also controlled by the westerlies and includes 892 valid trajectories. Path group3 (SYZR PM3, Bay of Bengal group) started from the Bay of Bengal and Arabian Sea and includes 1580 valid trajectories. Path group4 (SYZR PM4, northwest China group) corresponds with the northwest of China and includes 1331 valid trajectories. In PL events of the SYZR (Figure 4e,f), there are also four path groups. Path group1 (SYZR PL1), path group2 (SYZR PL2), path group3 (SYZR PL3), and path group4 (SYZR PL4) contain 937, 6413, 2990, and 2945 valid trajectories, respectively, similar to the origin of the SYZR PM paths.

The PM events of the SYR (Figure 5a–e) have five path groups. Path group1 (SYR PM1) started from a location much farther west than West Asia and includes 343 valid trajectories. Path group2 (SYR PM2) started from West Asia and includes 1247 valid trajectories. Path group3 (SYR PM3) started from the Bay of Bengal and Arabian Sea and includes 1865 valid trajectories. Path group4 (SYR PM4) corresponds with the northwest of China and includes 2272 valid trajectories. Path group5 (SYR PM5, southeast China group) corresponds with the east of China and includes 1702 valid trajectories. The PL events of the SYR (Figure 5f–h) also have five path groups. The five path groups (SYR PL1, SYR PL2, SYR PL3, SYR PL4, SYR PL5) contain 1627, 4785, 9831, 8030, and 8766 valid trajectories, respectively, similar to the origin of the SYR PM paths.

(a) (b)

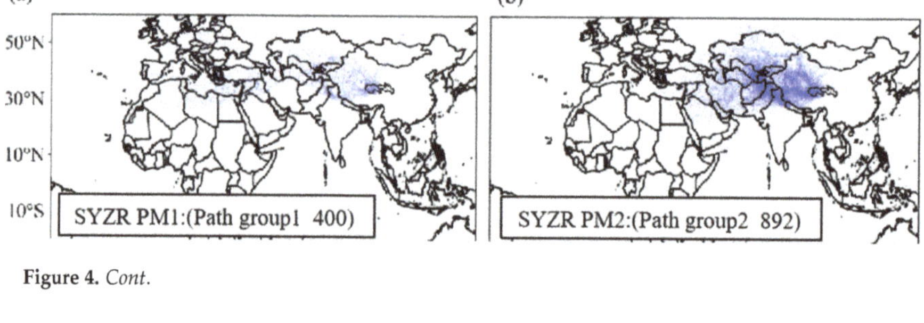

Figure 4. *Cont.*

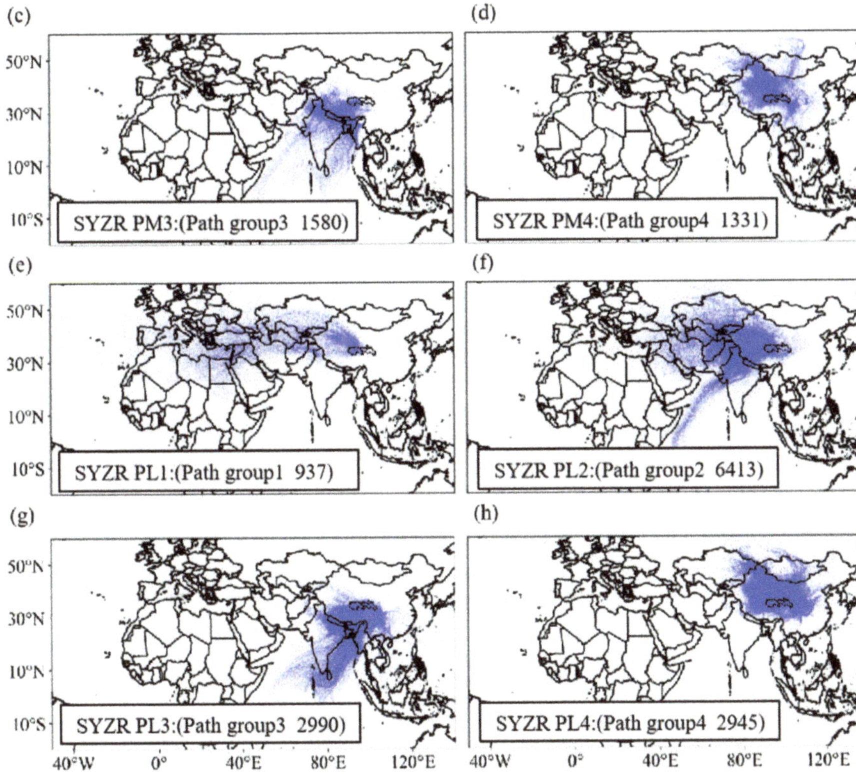

Figure 4. The clustering results of moisture trajectories for the SYZR.The blue lines represent trajectories. (**a–d**) are PM events; (**e–h**) are PL events. PM indicates the daily precipitation is greater than 10 mm events; PL indicates the daily precipitation is less than 10 mm events. SYR indicates the source regions of the Yellow River basin. SYZR indicates the source regions of the Yangtze River basin.

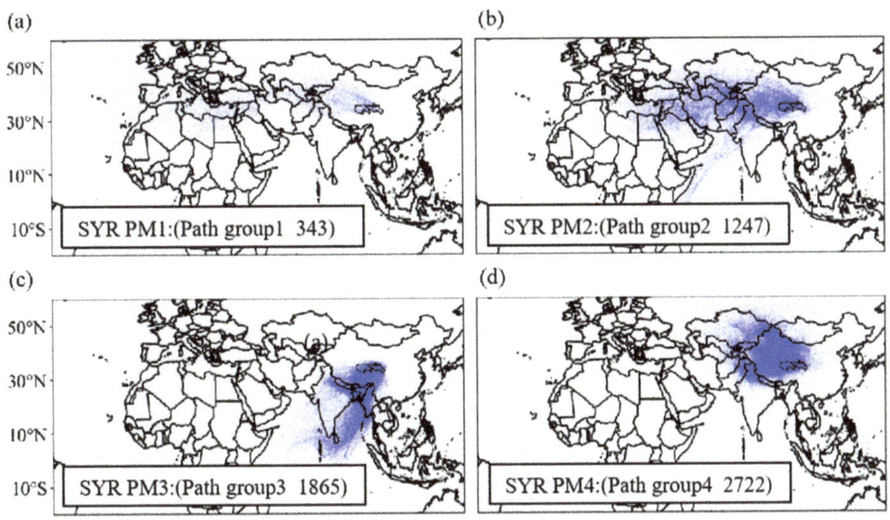

Figure 5. *Cont.*

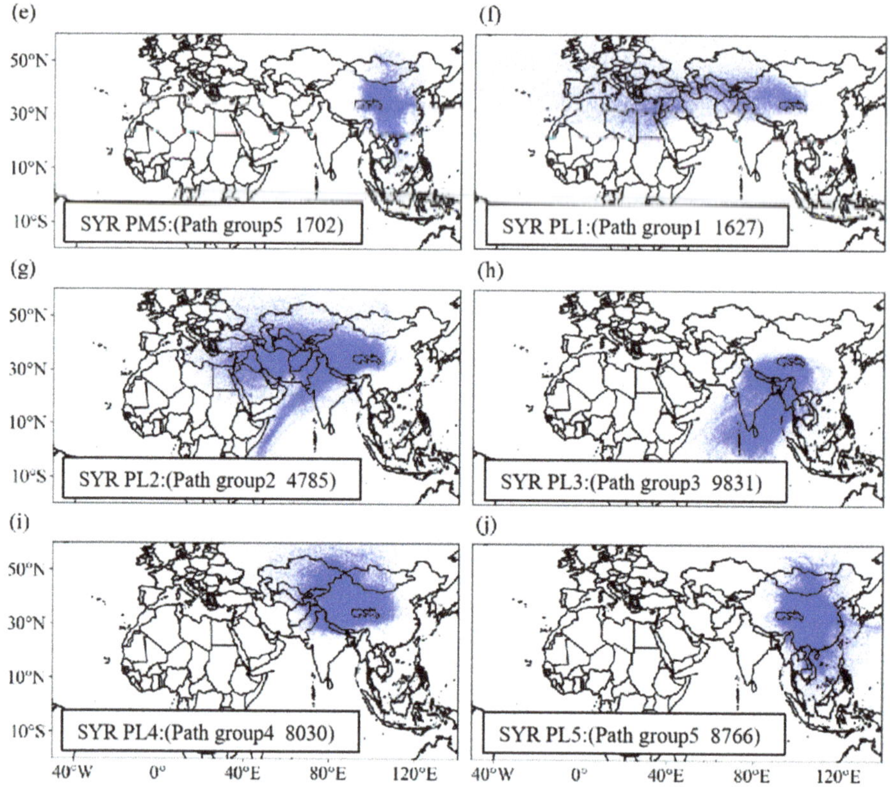

Figure 5. The clustering results of moisture trajectories for the SYR. The blue lines represent trajectories. (**a–e**) are PM events; (**f–j**) are PL events. PM indicates the daily precipitation is greater than 10 mm events; PL indicates the daily precipitation is less than 10 mm events. SYR indicates the source regions of the Yellow River basin. SYZR indicates the source regions of the Yangtze River basin.

3.2. Quantitative Importance of Major Moisture Transport Path Groups

Moisture undergoes multiple evaporation and precipitation losses during its transport, and specific humidity on the trajectories indicates changes in moisture content. The corresponding specific humidity changes along the paths of each cluster shown in Figures 6 and 7, indicate that many moisture paths undergo moisture losses before reaching the target region.

Overall, path group1 shows the most gradual humidity change, with lower specific humidity and the least contribution to precipitation in the river source regions. Path group3 exhibited the most drastic humidity changes, carrying a large amount of moisture from its origin and releasing a significant amount during transport. This path group contributed 2.5 times more to SYR precipitation than the SYZR. Path group2 has almost the same contribution rate to both river sources. Path group4 and path group5 show significant variations in moisture content along the trajectories.

Figure 6a–d for the SYZR during PM events in the last 6 h (0 ~ −6 h) show that the amount of moisture released by path group1, path group2, path group3, and path group4 is 90.7, 305.4, 870.8, and 813.6 $g{\cdot}kg^{-1}(6\ h)^{-1}$, respectively. These path groups contribute 4.36%, 14.68%, 41.87%, and 39.12%, respectively. Figure 6e–h during PL events for the SYZR in the last 6 h reveal that the amount of moisture released by path group1, path group2, path group3, and path group4 is 198.4, 995.2, 1930.1, and 1548.5 $g{\cdot}kg^{-1}(6\ h)^{-1}$, respectively. These path groups contributed 4.25%, 21.31%, 41.33%, and 33.16%, respectively.

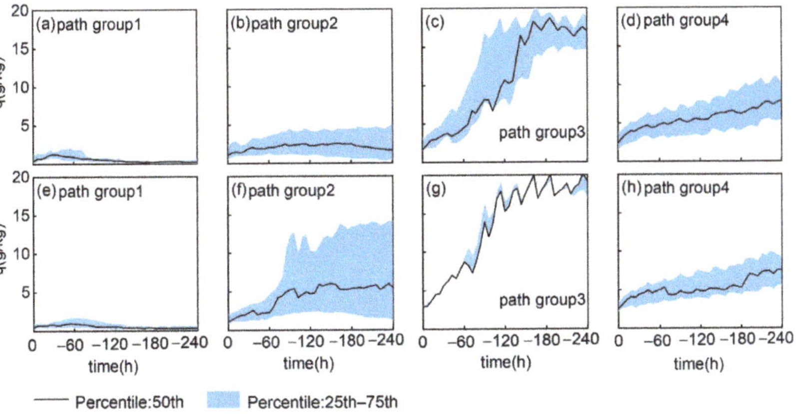

Figure 6. The specific humidity of different moisture path groups for the SYZR. (**a**–**d**) PM events of the SYZR; (**e**–**h**) PL events of the SYZR.

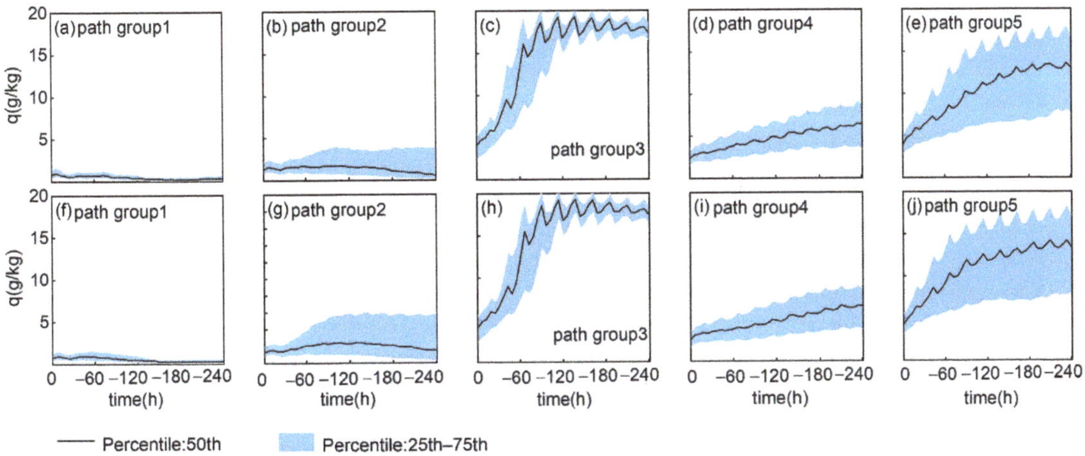

Figure 7. The specific humidity of different moisture path groups for the SYR. (**a**–**e**) PM events of the SYR; (**f**–**j**) PL events of the SYR.

For the SYR during PM events in the last 6 h as shown in Figure 7a–e, the amount of moisture released by path group1, path group2, path group3, path group4, and path group5 is 73.3, 312.5, 1050.7, 1145.8, and 941.5 g·kg^{-1}(6 h)$^{-1}$. These path groups contribute 2.08%, 8.87%, 29.83%, 32.53%, and 26.73%, respectively. Also, the amount of moisture released by path group1, path group2, path group3, path group4, and path group5 for the SYR during PL events in the last 6 h are shown in Figure 7f–j where 328.1, 1276.3, 6096.8, 6330.3, and 3470.5 g·kg^{-1}(6 h)$^{-1}$, respectively, were released. These path groups contributed 1.88%, 7.29%, 34.84%, 36.18%, and 19.83%, respectively.

In Figure 8, the total moisture released during PM and PL events for the SYZR was 2079.8 g·kg^{-1} (6 h)$^{-1}$ and 4669.8 g·kg^{-1} (6 h)$^{-1}$, respectively, while the total moisture released during PM and PL events for the SYR was 3522.0 g^{-1} (6 h)$^{-1}$ and 17,497.6 g·kg^{-1} (6 h)$^{-1}$, respectively. The PL events of the SYR released a more significant amount of moisture, which was 3.7 times higher than that released by the PL events of the SYZR. The Bay of Bengal group, northwest China group, and southeast China group (path group3, 4, and 5) transported more moisture for the SYR.

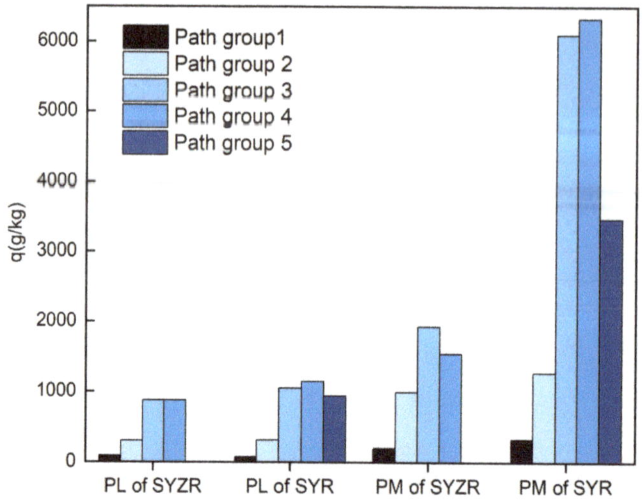

Figure 8. The amount of moisture released [$g{\cdot}kg^{-1}$] during different path groups in the last 6 h. PM of SYR indicates the PM events of the SYR. PL of SYR indicates the PL events of the SYR. PM of SYZR indicates the PM events of the SYZR. PL of SYZR indicates the PL events of the SYZR.

The statistical analysis of valid trajectories at different levels in the target region, as shown in Figure 9, reveals that the moisture release ($0 \sim -6$ h) heights in the river source regions are concentrated between 1500 and 2500 m above ground level (AGL). This altitude range is advantageous for conducting artificial weather modification operations.

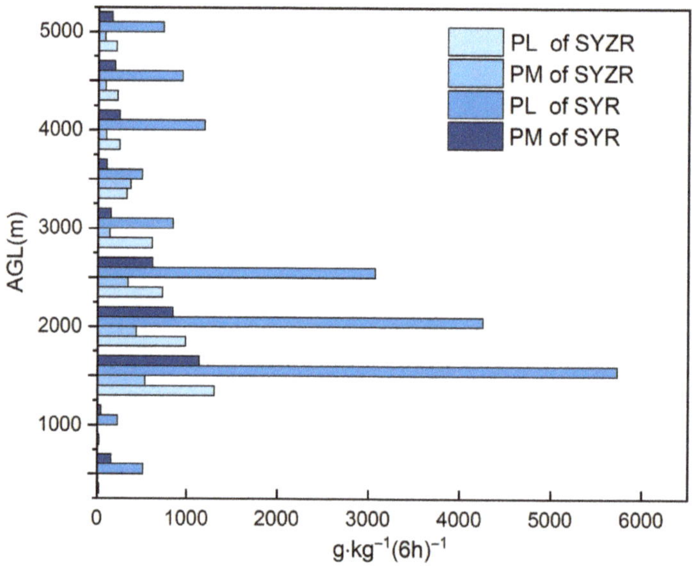

Figure 9. The amount of moisture released [g/kg] during different altitudes at the target region in the last 6 h. PM of SYR indicates the PM events of the SYR. PL of SYR indicates the PL events of the SYR. PM of SYZR indicates the PM events of the SYZR. PL of SYZR indicates the PL events of the SYZR.

3.3. Analysis of Moisture Exchange between the SYZR and SYR

The SYZR and SYR share valid trajectories that cross their respective borders because of their close proximities. Figure 10 presents results on the moisture exchange between the SYZR and SYR. During PM and PL events, the number of trajectories entering the SYZR from the SYR were 15 and 58, respectively, with a moisture release of about 7.1 g/kg and 16.30 g/kg. These values account for 0.34% and 0.35% of the total moisture release for the SYZR; these trajectories mainly belong to path group4 (northwest China group). Conversely, the number of trajectories entering the SYR from the SYZR were 798 and 3044 during the PM and PL events, with moisture releases of 257.4 g/kg and 1139.9 g/kg, constituting about 7.31% and 6.51% of its total moisture release, respectively; these trajectories mainly belong to path group3 (Bay of Bengal group), path group4 (northwest China group), and path group5 (southeast China group).

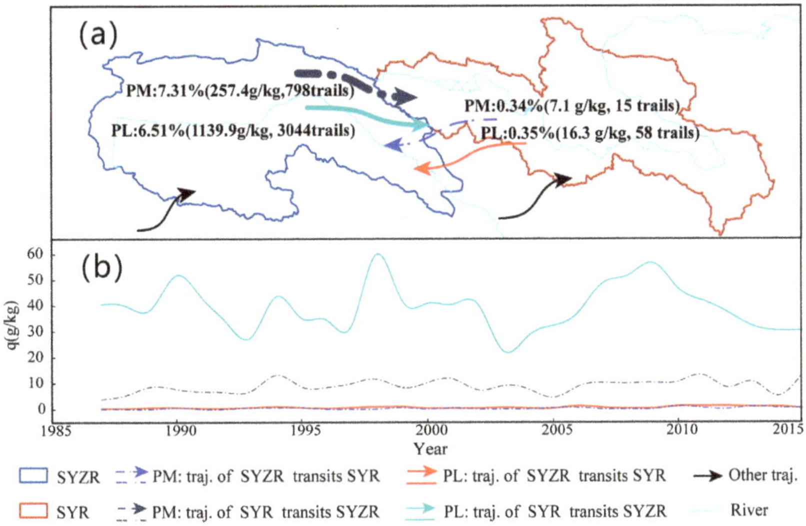

Figure 10. Moisture exchange between SYZR and SYR. (**a**) Total moisture exchange between SYZR and SYR from 1985 to 2015. (**b**) Annual variations in moisture exchange between SYZR and SYR.

The time series of the moisture exchange between the SYZR and SYR shows the annual summer moisture releases from the SYR to SYZR are close to zero but which from the SYZR to the SYR in PL events are about 39.26 g/kg (104.97 trajectories) and which from the SYZR to the SYR in PM events are about 8.84 g/kg (27.55 trajectories).

The attributed moisture release of each exchange trajectory between the SYZR and SYR is given in Figure 11. The moisture uptake locations belonging to each path group were divided into three categories: locations outside the source region of the two rivers (remote sources), locations inside the SYZR, and locations inside the SYR. For the PL events of the trajectories moving from the SYR to the SYZR, the moisture release of the four path groups is 1.37, 1.84, 0.28, and 9.34 $g{\cdot}kg^{-1}$ $(240\ h)^{-1}$, respectively. Path group1, path group2, and path group3 are of limited importance (0.29‰, 0.40‰, 0.06‰), while path group4 plays a leading role (2.01‰). For the PM events of the trajectories moving from the SYR to the SYZR, the moisture release of the four path groups is 0.46, 0.29, 0.00, and 5.78 $g{\cdot}kg^{-1}$ $(240\ h)^{-1}$, respectively. Path group1, path group2, and path group3 are of limited importance (0.22‰, 0.14‰, 0.00‰), while path group4 plays a leading role (2.77‰).

For the PL events of the trajectories moving from the SYZR to the SYR, the moisture release of the five path groups is 58.26, 54.58, 233.15, 242.48, and 344.56 $g{\cdot}kg^{-1}$ $(240\ h)^{-1}$, respectively. Path group1 and path group2 are of limited importance (3.33‰, 3.12‰), while path group3, path group4, and path group5 play leading roles (13.32‰, 13.85‰, 19.68‰).

For the PM events of the trajectories moving from the SYZR to the SYR, the moisture release of the five path groups is 11.26, 109.73, 31.88[1], 49.13, and 11.25 g·kg^{-1} (240 h)$^{-1}$, respectively. Path group1, path group3, and path group5 are of limited importance (3.12‰, 9.05‰, and 3.19‰), while path group2 and path group4 play leading roles (31.62‰ and 13.95‰).

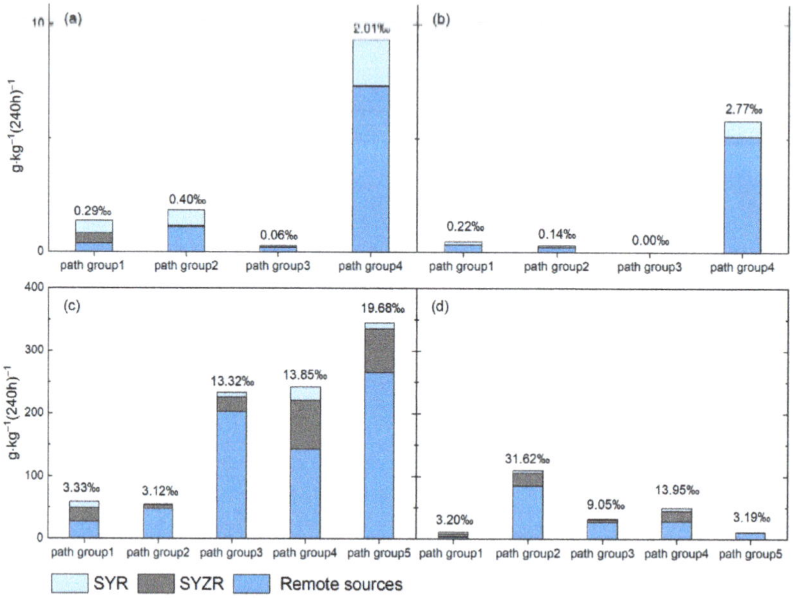

Figure 11. The amount of attributed moisture release [g·kg^{-1} (240 h)$^{-1}$] by locations of each path group responsible for (**a**) trajectories to the SYZR that transit the SYR during PL events, (**b**) trajectories to the SYZR that transit the SYR during PM events, (**c**) trajectories to the SYR that transit the SYZR during PL events, and (**d**) trajectories to the SYR that transit the SYZR during PM events. The term "remote sources" refers to the aggregate moisture release attributed to regions beyond the moisture uptake locations in the source region of the two rivers. Bars in dark blue represent the amount of the attributed moisture release of remote sources' moisture uptake locations, bars in gray represent the amount of the attributed moisture release of the SYR's moisture uptake locations, and bars in light blue represent the amount of the attributed moisture release of the SYZR's moisture uptake locations.

Given the higher proportion of trajectories from the SYZR entering the SYR, especially during the two events, it is crucial to analyze their trajectory frequency and average humidity (Figure 12). Its significant scientific and practical value of water resource security in the YR basin can be used for identifying the area where valid trajectories from the SYZR cross into the SYR.

The moisture paths' frequencies of different precipitation types are shown in Figure 12a,b. During the PM events of the SYR, regions with a high frequency of trajectories crossing from the SYZR are found in the northernmost part of India. At the same time, those for the PL are primarily located along the southern foothills of the Himalayas, extending through Tibet to the target area. The trajectories entering the SYZR are concentrated along moisture transport path group2, 3, and 4 (nearer west group, Bay of Bengal group, northwest China group) of the SYZR. The high-frequency areas of SYR moisture crossing into the SYZR are located in the southeastern parts of Zhiduo County and Zaduo County.

Figure 12c,d show the mean specific humidity of moisture paths from the SYRZ to the SYR for each grid cell. The average humidity of valid trajectories crossing from the SYZR into the SYR is lower in the TP, forming a band of low average humidity to the west of 34°

N latitude in the regions west of both the SYR and SYZR. Within the SYZR, the regions with a higher average humidity along the crossing trajectories are Chengduo County and Yushu City.

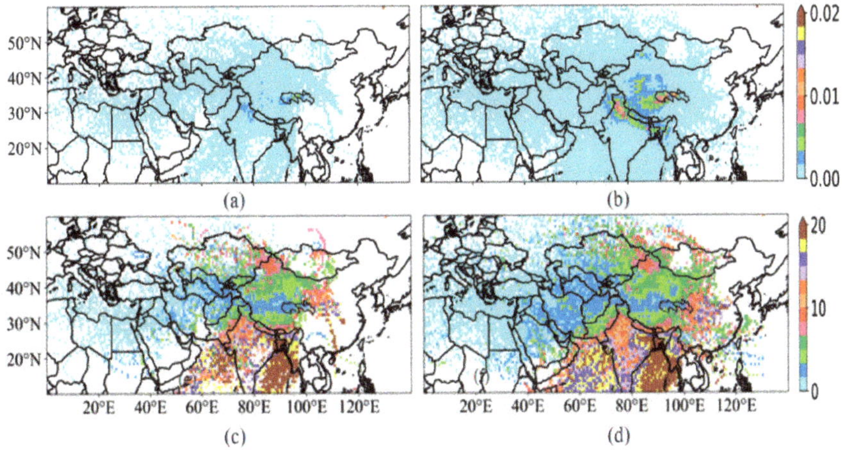

Figure 12. The frequency of the moisture paths (**above**) and mean specific humidity (g kg^{-1}) (**below**) from the SYZR to the SYR of moisture in (**a**,**c**) PM and (**b**,**d**) PL events.

The comparison reveals that although the areas with high trajectory frequency do not overlap with those of high average humidity, they are adjacent.

4. Conclusions and Discussion

4.1. Conclusions

This study investigates the major moisture transport pathways associated with two distinct types of daily precipitation occurrences (PM and PL events) throughout the summer in the SYZR and SYR regions. Additionally, it explores the characteristics of the reciprocal moisture exchange between the SYZR and the SYR.

The results of the moisture transport pathway identification suggest that path group3 (Bay of Bengal group) and path group4 (northwest China group) are the main and stable moisture transport pathways during PM and PL events in the SYZR [precipitation contribution: path group3, 41.87% (PM), 41.33% (PL); path group4, 39.12% (PM), 33.16% (PL)]. Path group3 (Bay of Bengal group), path group4 (northwest China group), and path group5 (southeast China group) are the main and stable moisture pathways during PM and PL events in the SYR [precipitation contribution: path group3, 29.83% (PM), 34.84% (PL); path group4, 32.53% (PM), 36.18% (PL); path group5, 26.73% (PM), 19.83% (PL)].

The amount of moisture transiting from the SYZR to the SYR was about 60 times that of the reverse, accounting for approximately 6.9% of the total precipitation moisture released in the SYR. The moisture pathways transiting from the SYZR to the SYR were present from 1986 to 2015. The amount of moisture through path group5 (southeast China group) from the SYZR to the SYR is relatively larger than other path groups during PL events. The amount of moisture through path group2 (nearer west group) from the SYZR to the SYR is relatively larger than other path groups during PM events. The areas with high trajectory frequency from the SYZR to the SYR were the southeast of Zhiduo County and the southeast of Zaduo County.

4.2. Discussion

Atmospheric water resources are abundant, such as a huge renewable reservoir [29]. This paper analyzed the characteristics of the moisture transport between the SYZR and

SYR, which is of great value to provide a full picture of the atmospheric hydrological cycle of the SYZR and SYR. There have been numerous experimental studies on the development of atmospheric water resources through artificial rain enhancement operations within the basin [30]. The study of moisture exchange between the SYZR and SYR provides a new perspective for the diversified development and utilization of atmospheric water resources in subsequent adjacent basins; this issue deserves a systematic and in-depth study.

As for the moisture pathways for the SYZR and SYR, there are already some studies on the moisture pathways of a related region. Yu Zhang et al. [13] noted that moisture paths from the northwest covering northwest China and Central Asia and moisture paths from southern and southeastern China are stable moisture transport pathways during extreme precipitation and moderate precipitation events over the TRHR of the TP. Nepiao Xu [12] noted that the climatological moisture sources over the TP were identified mainly at the southern and western boundaries, with the moisture sources at the southern boundaries originating from the Arabian Sea and Bay of Bengal and the moisture sources at the western boundary being transported by mid-latitude westerlies. Rong Liu [31] noted that the moisture pathways mainly came from the land and ocean in the west during the extreme drought episodes based on the standardized precipitation index (SPI). When the SPI indicated an extremely wet state, the transport mainly occurred to the east and south, including the Pacific Ocean and the South China Sea, where moisture entered the mainland from East/South China, passed through the Yangtze River basin, and finally reached the TRHR from the east or south of the TP. This is consistent with the view in this paper that the main moisture pathways are path group3 (Bay of Bengal group), path group4 (northwest China group), and path group5 (southeast China group).

The limitations of this study are that the K-means cluster results in the present study are relatively conservative; short paths tend to be grouped together as a clump with unclear directionality [such as path group1 (farther west group) and path group2 (nearer west group)], which is also noted by Borge et al. [32]. We think that this problem can be alleviated by introducing the direction of the previous part of the trajectories and consider it worthy of further efforts. Furthermore, the accuracy of the simulation results is influenced by various factors. Firstly, the influence of the data on simulation results is crucial. NOAA provides various data compatible with the HYSPLIT model, including NNR1, GDAS, and NCEP, among others. However, most studies rely on a single data source. For instance, Zhang et al. used NNR1 data to simulate the moisture trajectories of precipitation in the TRHR, while Xu et al. utilized ERA-interim data to simulate the moisture trajectories of precipitation in the Qinghai–Tibet Plateau. On the other hand, both Ayantobo et al. and our study employed NCEP data to simulate the moisture trajectories of precipitation. This reliance on a single data source may introduce biases in the results, and the reliability of the data is difficult to ascertain. Secondly, the settings for simulation altitude and starting time in the HYSPLIT model can also affect the accuracy of the results. Presently, most studies analyzing the characteristics of moisture transport simulate the moisture trajectories of daily precipitation events within the study area. Some studies only simulate trajectories at a single altitude on the day of precipitation, while others simulate trajectories at multiple altitudes on the day of precipitation, and still others simulate trajectories at multiple altitude levels until the end of the precipitation day. To refine the simulation, precipitation can be divided into events, and the starting point of the trajectory simulation can be set as the endpoint time of each precipitation event. This approach helps avoid inaccuracies in the choice of the simulation starting time. Additionally, the selection of simulation altitudes should consider the range of altitudes at which moisture is predominantly transported and should be based on the predominant distribution of moisture flux heights in the study area.

Through the analysis of moisture exchange characteristics between the two river source regions, it has been found that the moisture from the SYZR to the SYR accounts for 6.9% of the precipitation moisture in the SYR. Although the quantity is relatively small, it indicates the existence of a channel for moisture exchange. Regions such as Zhiduo County and the southeastern part of Zado County are characterized by high-frequency trajectories of moisture exchange. In future research, advanced atmospheric simulation and prediction models, such as WRF, will be utilized to simulate artificial weather modification operations in key areas of moisture exchange (high-frequency trajectory regions) between the two river source regions. The impact on precipitation in the two river source regions will be evaluated. This will further analyze the feasibility of achieving the cross-basin allocation of atmospheric water resources and enhance the understanding of airborne water resource management.

Author Contributions: Conceptualization, J.W., B.K. and O.O.A.; methodology, J.W. and B.K.; software, B.K.; validation, J.W., B.K. and O.O.A.; formal analysis, J.W., B.K., H.Y. and O.O.A.; data curation, B.K.; writing—original draft preparation, B.K.; writing—review and editing, O.O.A., J.W., B.K. and H.Y.; project administration, J.W.; funding acquisition, J.W. All authors have read and agreed to the published version of the manuscript.

Funding: This research was funded by the National Natural Science Foundation of China, grant number U2243232; the Major Science and Technology Project of Qinghai Province, grant number 2021-SF-A6; and Qinghai University Young and Middle-aged Scientific Research Fund, grant number 2020-QGY-13.

Institutional Review Board Statement: Not applicable.

Informed Consent Statement: Not applicable.

Data Availability Statement: The data presented in this study are available on request from the corresponding author. The data are not publicly available due to approval process.

Conflicts of Interest: The authors declare no conflicts of interest.

Appendix A

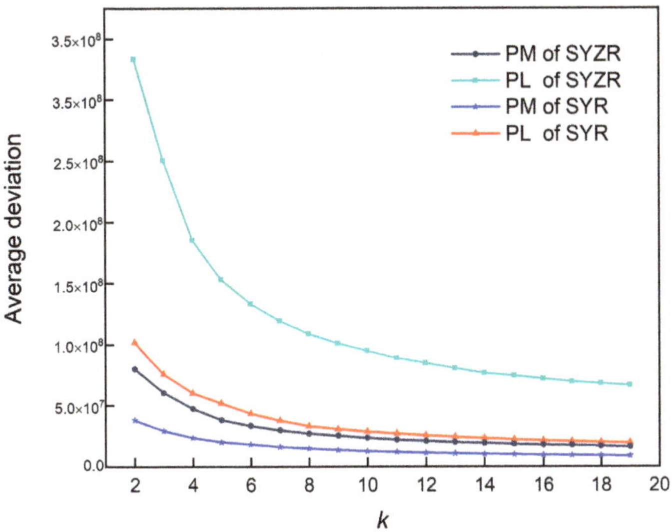

Figure A1. The average deviation for different numbers of clusters. PM of SYR indicates the PM events of the SYR. PL of SYR indicates the PL events of the SYR. PM of SYZR indicates the PM events of the SYZR. PL of SYZR indicates the PL events of the SYZR. PM indicates the daily precipitation is greater than 10 mm events; PL indicates the daily precipitation is less than 10 mm events. SYR indicates the source regions of the Yellow River basin. SYZR indicates the source regions of the Yangtze River basin. *K* indicates the clustering number.

Table A1. Silhouette coefficients for different numbers of clusters.

Region	Event	$K = 2$	$K = 3$	$K = 4$	$K = 5$	$K = 6$	$K = 7$	$K = 8$
SYZR	PM	0.513	0.295	0.320	0.313	0.313	0.311	0.314
	PL	0.587	0.410	0.433	0.427	0.461	0.472	0.477
SYR	PM	0.490	0.332	0.361	0.368	0.365	0.355	0.356
	PL	0.488	0.328	0.374	0.376	0.373	0.353	0.357

K indicates the clustering number.

References

1. Wang, Y.; Ding, Y.; Ye, B.; Liu, F.; Wang, J.; Wang, J. Contributions of climate and human activities to changes in runoff of the Yellow and Yangtze rivers from 1950 to 2008. *Sci. China Earth Sci.* **2012**, *56*, 1398–1412. [CrossRef]
2. Ayantobo, O.O.; Wei, J.; Shi, Y.; Wang, G. Climatological changes in rainfall distributions at different rain-rates under Qinghai-Tibet Plateau warming during 1981–2060. *Theor. Appl. Climatol.* **2023**, *152*, 663–679. [CrossRef]
3. Wei, J. *A Research Report on Sanjiangyuanecological Protection: Volume of Hydrology and Water Resources*; Social Sciences Academic Press: Beijing, China, 2017; p. 2.
4. Lu, W.; Wang, W.; Shao, Q.; Yu, Z.; Hao, Z.; Xing, W.; Yong, B.; Li, J. Hydrological projections of future climate change over the source region of Yellow River and Yangtze River in the Tibetan Plateau: A comprehensive assessment by coupling RegCM4 and VIC model. *Hydrol. Process.* **2018**, *32*, 2096–2117. [CrossRef]
5. Cao, L.; Pan, S. Changes in precipitation extremes over the "Three-River Headwaters" region, hinterland of the Tibetan Plateau, during 1960–2012. *Quat. Int.* **2014**, *321*, 105–115. [CrossRef]
6. Hu, Y.; Zhou, Y.; Wang, Y.; Lu, F.; Xiao, W.; Hou, B.; Yu, Y.; Liu, J.; Xue, W. Impacts of Precipitation Type Variations on Runoff Changes in the Source Regions of the Yangtze and Yellow River Basins in the Past 40 Years. *Water* **2022**, *14*, 4115. [CrossRef]
7. Tang, Q.; Liu, Y.; Zhang, C.; Su, F.; Li, Y.; Gao, Y.; Li, W.; Chen, D. Research progress on moisture source change of precipitation over the Tibetan Plateau and its surrounding areas. *Trans. Atmos. Sci.* **2020**, *43*, 1002–1009. [CrossRef]
8. Li, H.; Pan, X. An overview of research methods on water vapor transport and sources in the Tibetan Plateau. *Adv. Earth Sci.* **2022**, *37*, 1025–1036.
9. Shi, X.; Shi, X. Climatic characteristics of water vapor budget over southeastern Tibetan Plateau in summer and their effects. *JAMS* **2008**, *19*, 41–46.
10. Qiang, A.; Wang, N.; Xie, J.; Wei, J. Analysis of Water Vapor Change and Precipitation Conversion Efficiency Based on HYSPLIT Backward Trajectory Model over the Three-River Headwaters Region. *J. Coast. Res.* **2020**, *105*, 6–11. [CrossRef]
11. Ayantobo, O.O.; Wei, J.; Hou, M.; Xu, J.; Wang, G. Characterizing potential sources and transport pathways of intense moisture during extreme precipitation events over the Tibetan Plateau. *J. Hydrol.* **2022**, *615*, 128734. [CrossRef]
12. Xu, N.; Zhong, L.; Ma, Y.; Zou, M.; Huang, Z. A study on the water vapor transport trend and water vapor source of the Tibetan Plateau. *Theor. Appl. Climatol.* **2020**, *140*, 1031–1042. [CrossRef]
13. Zhang, Y.; Huang, W.; Zhong, D. Major Moisture Pathways and Their Importance to Rainy Season Precipitation over the Sanjiangyuan Region of the Tibetan Plateau. *J. Clim.* **2019**, *32*, 6837–6857. [CrossRef]
14. Tang, Q.; Zhang, C.; Chen, D. Recent Changes in the Moisture Source of Precipitation over the Tibetan Plateau. *J. Clim.* **2017**, *30*, 1807–1819. [CrossRef]
15. Zhu, L.; Liu, R.; Wang, X.; Wang, Z.; Wen, J.; Zhao, Y.; Xie, Y.; Zhang, T. The characteristics of the water vapor transport and associated sources under abnormal precipitation conditions in the source region of the Yellow River using FLEXPART. *Plateau Meteorol.* **2019**, *38*, 484–496. [CrossRef]
16. Li, J.; Li, T.; Wang, G.; Wei, J.; Zhong, D.; Su, Y.; Fu, X. Atmospheric water resource and precipitation conversion. *Chin. Sci. Bull.* **2018**, *63*, 2785–2796. [CrossRef]
17. Ren, Z.H.; Zou, F.L.; Yu, Y.; Wang, G.A.; Zhang, Z.F.; Fan, S.H.; Zhang, Z.Q.; Sun, C. Dataset of Daily Climate Data from Chinese Surface Stations, Version 3.0. National Meteorological Information Center. Available online: https://data.cma.cn/data/cdcindex/cid/f0fb4b55508804ca.html (accessed on 5 January 2018).
18. *GB/T 28592-2012 Grade of Precipitation*; General Administration of Quality Super Vision, Inspection and Quarantine of the People and Republic of China. Standards Press of China: Beijing, China, 2012.
19. Yatagai, A.; Namiguchi, N.; Arakawa, O.; Hamada, A.; Yasutomi, N.; Nitoh, A. APHRODITE: Constructing a Long-Term Daily Gridded Precipitation Dataset for Asia Based on a Dense Network of Rain Gauges. *Bull. Am. Meteorl. Soc.* **2012**, *93*, 1401–1415. [CrossRef]
20. Draxler, R.R.; Hess, G.D. *Description of the HYSPLIT_4 Modeling System*; NOAA Technical Memorandum ERL ARL-224; National Oceanic and Atmospheric Administration: Washington, DC, USA, 1997; 224p.
21. Draxler, R.R.; Hess, G.D. An overview of the HYSPLIT_4 modelling system for trajectories, dispersion, and deposition. *Aust. Meteorol. Mag.* **1998**, *47*, 295–308.
22. Wagner, T.; Andreae, M.O.; Beirle, S.; Dörner, S.; Mies, N.; Shaiganfar, R. MAX-DOAS observations of the total atmospheric water vapour column and comparison with independent observations. *Atmos. Meas. Tech.* **2013**, *6*, 131–149. [CrossRef]

23. Trenberth, N.E. Atmospheric Moisture Recycling: Role of Advection and Local Evaporation. *J. Clim.* **1999**, *12*, 1368–1381. [CrossRef]
24. Trenberth, N.E. Atmospheric Moisture Residence Times and Cycling: Implications for Rainfall Rates and Climate Change. *Clim. Chang.* **1998**, *39*, 667–694. [CrossRef]
25. Cohen, M.D.; Stunder, B.J.B.; Rolph, G.D.; Draxler, R.R.; Stein, A.F.; Ngan, F. NOAA's HYSPLIT Atmospheric Transport and Dispersion Modeling System. *Bull. Am. Meteorol. Soc.* **2015**, *96*, 2059–2077. [CrossRef]
26. Rousseeuw, P.J. Silhouettes: A graphical aid to the interpretation and validation of cluster analysis. *J. Comput. Appl. Math.* **1987**, *20*, 53–65. [CrossRef]
27. Sodemann, H.; Schwierz, C.; Wernli, H. Interannual variability of Greenland winter precipitation sources: Lagrangian moisture diagnostic and North Atlantic Oscillation influence. *J. Geophys. Res.* **2008**, *113*, D03107. [CrossRef]
28. Sun, B.; Wang, H. Moisture Sources of Semiarid Grassland in China Using the Lagrangian Particle Model FLEXPART. *J. Clim.* **2014**, *27*, 2457–2474. [CrossRef]
29. Wahlgren, R.V. Atmospheric water vapour processor designs for potable water production: A review. *Water Res* **2001**, *35*, 1–22. [CrossRef]
30. Flossmann, A.I.; Manton, M.; Abshaev, A.; Bruintjes, R.; Murakami, M.; Prabhakaran, T.; Yao, Z. Review of Advances in Precipitation Enhancement Research. *Bull. Am. Meteorol. Soc.* **2019**, *100*, 1465–1480. [CrossRef]
31. Liu, R.; Wang, X.; Wang, Z. Atmospheric moisture sources of drought and wet events during 1979–2019 in the Three-River Source Region, Qinghai-Tibetan Plateau. *Theor. Appl. Climatol.* **2022**, *149*, 487–499. [CrossRef]
32. Borge, R.; Lumbreras, J.; Vardoulakis, S.; Nassomenos, P.; Rodriguez, E. Analysis of long-range transport influences on urban PM10 using two-stage atmospheric trajectory clusters. *Atmos. Environ.* **2007**, *41*, 4434–4450. [CrossRef]

atmosphere

MDPI

Article

Spatio-Temporal Changes of Vegetation Net Primary Productivity and Its Driving Factors on the Tibetan Plateau from 1979 to 2018

Mingwang Li [1], Qiong Li [2,*] and Mingxing Xue [1]

[1] College of Water Resources & Civil Engineering, Qinghai University, Xining 810016, China; ys210859000404@qhu.edu.cn (M.L.); ys210859000366@qhu.edu.cn (M.X.)
[2] State Key Laboratory of Plateau Ecology and Agriculture, Qinghai University, Xining 810016, China
* Correspondence: liqiong_skl@qhu.edu.cn

Abstract: The Net Primary Productivity (NPP) of the Tibetan Plateau (TP) has undergone significant changes since the 1980s. The investigation of the spatiotemporal changes of NPP and its driving factors is of significant importance. Here, we analyze the spatial and temporal trends of Net Primary Production (NPP) and the effects of meteorological factors on the NPP change on the Tibetan Plateau (TP) using version 5.0 of the Community Land Model. The results showed that the average NPP was 256 (g $C{\cdot}m^2{\cdot}yr^{-1}$) over the past 40 years, with a continuously increasing trend of 2.38 (g $C{\cdot}m^2{\cdot}yr^{-1}$). Precipitation was the main factor affecting NPP changes, temperature had no significant effect on NPP changes, while radiation showed a negative trend. Changes in precipitation, temperature and radiation account for approximately 91%, 5.3%, and 3.8% of NPP variation, respectively. Based on grass coverage, we categorized alpine grasslands into three types: high, medium, and low coverage. Our findings indicate the NPP change of the high-coverage grasslands was mainly affected by precipitation, and then the temperature and radiation. Comparatively, the precipitation change is the driving factor of the increased NPP of low-coverage grasslands, but the temperature increase is the negative factor. Our studies have implications for assessing and predicting vegetation responses to future climate change.

Keywords: NPP; CLM 5.0 model; Tibetan Plateau; spatio-temporal dynamic; climate factor

1. Introduction

The Tibetan Plateau (TP) is the highest plateau in the world with an average altitude of about 4000 m and is known as the "Roof of the World" [1]. In recent decades, temperature increased significantly on the TP, and TP warming is occurring twice as fast as global warming [2]. The unique high and cold ecosystem of the TP has a simple ecological structure and is very vulnerable to climate change [3]. Net Primary Productivity (NPP) of vegetation is a key ecological indicator for detecting C fluxes, and it refers to the amount of organic carbon produced by plant communities in an ecosystem per unit of time and area [4]. NPP is the remaining portion of the total amount of organic matter fixed by photosynthesis minus the amount consumed by its own respiration. It is an important indicator of ecosystem health and ecological balance [5], as well as global carbon balance [6].

Many models have been conducted to calculate and analyze the change of NPP and its influencing factors. The field measurement is the direct method to investigate the NPP. Given the harsh environmental conditions of the TP, field measurement data have limited spatial coverage; hence it is often used as the verification data. Statistical models mainly include traditional climate correlation models and remote sensing data-driven models [7]. Statistical models were simple in structure [8–10], but these models lack the theoretical foundation of plant ecophysiology as well as the ability to illustrate how the meteorological environment interacts with the ecosystem [11]. The potential of the machine learning method has been widely acknowledged in the field of ecological science in recent years [12,13]. The machine learning approaches (e.g., Random Forest, RF; Support Vector Machines, SVM; k-Nearest Neighbor, KNN) have been successfully used to derive carbon fluxes using the fluxnet measurements and satellite-derived variables [14–17]. In addition, machine learning algorithms such as gradient boosting (GB) [18], extreme gradient boosting (XGB) [19], and light gradient boosting machine (LGBM) [20], have achieved good performance in carbon flux retrieval. These algorithms are under the gradient boosting framework and can provide parallel tree boosting to derive carbon fluxes in a fast and accurate way.

The process-based models combined with the physiological characteristics of plant growth and development were able to simulate the functional processes within the ecosystem and construct the interaction process between the plant interior and external meteorological conditions. Process-based models are represented by the CLM model [21,22], Biome-BGC model [23], and TEM model [24], etc. The Community Land Model (CLM) is the most developed and widely used land surface model in the world and has been called the largest terrestrial biosphere model [25,26], also suitable for the Tibetan Plateau. Lin et al. [27] used the CLM4.5 model to simulate the spatial and temporal variations of water use efficiency (WUE) in the Tibetan Plateau, and its response to environmental factors found that WUE showed an increasing trend and its main influencing factors were the CO_2 concentration and temperature; Hao et al. [28] used the CLM model assessed NPP and a variety of water-related ecosystem services in the Qinghai Plateau region from 2000 to 2018, revealing that vegetation dominated the relative locations of NPP thresholds.

Precipitation, temperature, nitrogen deposition, carbon dioxide concentration, and many other factors can affect the NPP of vegetation [29]. Tu et al. [30] used correlation analysis combined with the Thornthwaite Memorial model to analyze the changes in the spatial distribution pattern of NPP in China from 2000 to 2017. They suggested that NPP is mainly affected by temperature in the eastern Tibetan Plateau. Zha et al. [31] used the CASA model and biased correlation analysis to analyze the NPP change and the influence factors on the Tibetan Plateau. They concluded that the controlling factor of NPP change shifted from the temperature during 2000–2009 to the precipitation during 2010–2019. Zheng et al. [32] used the CASA model and correlation analysis to analyze the NPP change on the Tibetan Plateau from 2001 to 2015, and suggested that the effect of the solar radiation was greater than the temperature and precipitation to the NPP change of the TP. Qiu et al. [33] used CESM 1.0.4 to explore the impact of the different concentrations of CO_2 and climate change on the vegetation on the TP. They suggested that the increasing CO_2 concentration increased the values of GPP, NPP and other parameters, but climate change had a negative impact on the vegetation structure and productivity in the TP.

Previous research analyzed the response of NPP to the meteorological factor, but it can be seen that obvious contradictions and controversies existed in these researches. As one of the most powerful models with the strongest simulation capability for land surface, CLM is well suited for the simulation of NPP. In particular, the latest generation of the CLM 5.0 model improves the accuracy of soil hydrothermal simulation in the permafrost zone, which makes it more suitable for the Tibetan Plateau. The objective of this study is twofold: (1) to investigate the spatial and temporal trends of NPP and quantify the effects of meteorological factors on the NPP in the Tibetan Plateau and (2) to indicate differences in NPP changes and driving factors over the grasslands with different vegetation cover. This paper is organized as follows. The methods and data are introduced in Section 2. The main results are presented in Section 3. Discussion and conclusions are presented in Sections 4 and 5.

2. Materials and Methods

2.1. Study Area

The Tibetan Plateau (73–105° E, 25–40° N) is located in the western region of China, covering a total area of approximately 2.5×10^6 km^2, with an average elevation of 4500 m

(Figure 1). Grassland covers over 60% of the plateau, while forests and trees only occur in small areas in the south TP. The Tibetan Plateau is in the arid and semi-arid climate zone, with an average annual temperature below 0 °C and an average annual precipitation of approximately 500 mm. The regional ecology is fragile, with obvious geographical differences and distinct wet and dry seasons.

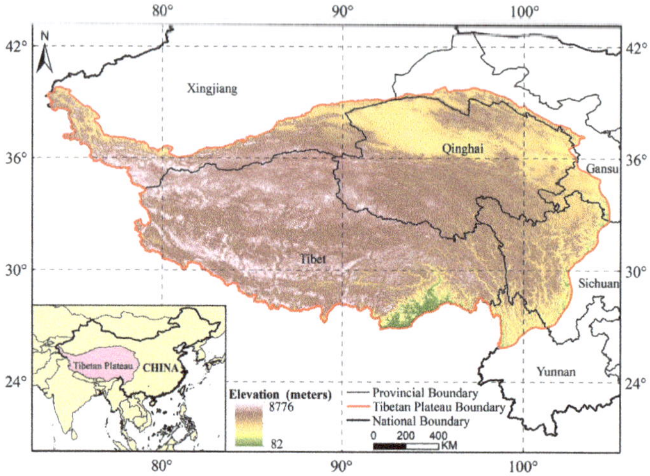

Figure 1. Geographic location map of the Tibetan Plateau.

2.2. Model Description and Data Sources

The Community Land Model (CLM), developed by the National Center for Atmospheric Research (NCAR) in the United States, is capable of simulating biogeophysical and biogeochemical processes, including radiative transfer, the vegetation–soil–hydrological cycle, the soil carbon–nitrogen cycle, and vegetation photosynthesis [34]. The Community Earth System Model (CESM) has been coupled with CLM models since version 4.0 and runs under the framework of CESM [35]. CLM uses a multi-layer nested subgrid hierarchy to represent the spatial heterogeneity of the land surface. Each grid cell consists of three sub-grids: land units, snow/soil columns, and plant functional types (PFTs). The number of land units can vary within each grid cell, with each land unit potentially containing a different number of snow/soil columns, and each column having multiple PFTs.

CLM 5.0 has made significant improvements compared to CLM 4.5 in various aspects including soil and plant hydrology, snow density, river modeling, crop modeling, carbon and nitrogen cycling, and coupling processes. In terms of hydrology, CLM 5.0 now features increased soil vertical resolution (20 soil layers + 5 bedrock layers) and spatially variable soil depth (ranging from 0.4 to 8.5 m), as well as the removal of unconfined aquifer and no flux lower boundary condition. In terms of vegetation dynamics, CLM 5.0 has enhanced the plant hydraulic stress model for water transport through vegetation by replacing empirical soil moisture stress formulation and introducing hydraulic redistribution. Furthermore, in terms of carbon and nitrogen cycling processes, CLM 5.0 replaces dynamic NPP-based allocation schemes with flexible plant C:N ratios to eliminate instantaneous down-regulation of photosynthesis based on mineral N availability [36]. Most importantly, CLM 5.0 improves the permafrost physical processes compared to other process-based models [37]. Thus, it can be seen that CLM 5.0 includes carbon feedback and improves the accuracy of soil water-heating process simulation, especially in the permafrost zone. Therefore, CLM 5.0 is widely used for simulating carbon, water, and energy exchange between the global and regional scale land and atmosphere [38–41], and it is well suited for simulating carbon, water and other variables in the Tibetan Plateau region [42,43].

In CLM 5.0, the NPP is simulated with the biogeochemistry (BGC) model. The NPP can be calculated by subtracting plant respiration from gross primary productivity (GPP). The detailed process and formula are explained in Lawrence et al. [36]. The NPP_{coh} is the cohort-specific Net Primary Productivity, which is calculated as follows:

$$NPP_{coh} = GPP_{coh} - R_{plant,coh} \tag{1}$$

GPP is calculated as follows:

$$GPP_{\text{coh}} = 12 \times 10^{-9} \sum_{z=1}^{nz(con)} gpp_{Cl,ft,z} A_{crown,coh} elai_{Cl,ft,z} \tag{2}$$

where the GPP_{coh} is cohort-specific gross primary productivity, $gpp_{Cl,ft,z}$ is the rate of gross photosynthesis, $A_{crown,coh}$ is the Crown area, and $elai_{Cl,ft,z}$ is the sum of exposed leaf area index. Plant respiration is calculated as follows:

$$R_{plant} = R_{g,coh} + R_{m,coh} \tag{3}$$

where the R_{plant} is the plant respiration, and $R_{g,coh}$ and $R_{m,coh}$ is the growth and maintenance respiration.

The simulation performance of land surface models can be influenced by several slow-changing parameters. The spin-up process is typically used to provide initial conditions for the model [44]. Spin-up makes the continuous adjustment of various model parameters, such as soil temperature and humidity, latent heat, and sensible heat, until the model reaches an equilibrium state [45]. This study uses the China Meteorological Forcing Dataset (CMFD) to complete the spin-up process for 400 years in order to ensure equilibrium.

The climate-forcing dataset used in this study is the CMFD. CMFD comprises seven near-surface meteorological variables: 2 m air temperature, surface pressure, near-surface specific humidity, 10 m wind speed, downward shortwave radiation, downward longwave radiation, and precipitation rate [46]. The dataset integrates meteorological observations and remote sensing/reanalysis data. The temporal and spatial resolution of the dataset is 3 h and 0.1°. These remote sensing/reanalysis data include Princeton reanalysis data, GLDAS data, GEWEX-SRB radiation data, and TRMM precipitation data [47,48]. CMFD is more accurate than other meteorological forcing datasets [49]. Land use data from 2020 were used in the study and were provided by the Data Center for Resources and Environmental Sciences, Chinese Academy of Sciences (RESDC) [50]. The workflow of this paper is shown in Figure 2 and detailed data information is shown in Table A1.

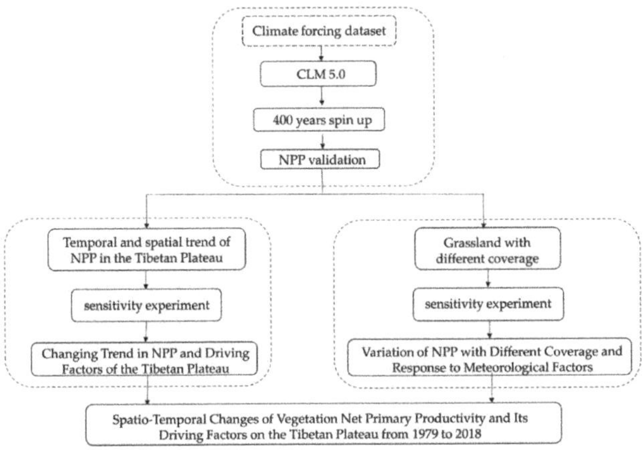

Figure 2. Workflow chart.

The measured NPP data collected from the published research are used as ground-truthing data to validate the NPP simulated by CLM 5.0 [51]. In addition, the spatial patterns of NPP simulation from the CLM5.0 were also validated using the NPP calculated from the Carnegie-Ames-Stanford Approach (CASA) model. The CASA model is known for its computational simplicity, operational feasibility, and parameter sharing options. It has greatly facilitated NPP simulations at both global and regional scales [52]. Currently, the CASA model has become one of the most effective methods to study and calculate large-scale NPP [53], it also suits the NPP simulation in the Tibetan Plateau region [54,55]. Given that the CASA model relies on MODIS NDVI data which starts from the year 2000, this study used the CASA model simulated NPP from the years 2000 to 2018 as validation data.

2.3. Experimental Design and Analytical Methods

This study conducted a series of control experiments and sensitivity experiments to investigate the influence of climate factors (temperature, precipitation, radiation) on the NPP (Table 1). Scenario one served as the control experiment, while scenarios two to four were sensitivity experiments. In scenario one, the CLM 5.0 model was driven by CMFD from 1979 to 2018 to simulate the NPP. In scenarios two to four, the CLM model was driven by different atmospheric forcing data. In scenario two, temperature was held constant and set by the average values from 1979 to 2018, while precipitation and radiation remained consistent with the control experiment. In scenario three, precipitation was held constant and set by the 1979–2018 average, while temperature and radiation remained consistent with the control experiment. In scenario four, radiation was held constant and set by the 1979–2018 average, while temperature and precipitation remained consistent with the control experiment. These scenarios were designed to investigate the impact of individual meteorological forcing factors on the NPP. By comparing the control experiment and sensitivity experiments, we can quantify the extent of the meteorological forcing factors to the NPP change on the plateau.

Table 1. Experimental design.

Experimental	Climate Factors		
	Temperature	Precipitation	Radiation
Scenario One	T	T	T
Scenario Two	C	T	T
Scenario Three	T	C	T
Scenario Four	T	T	C

Note: T represents parameters that change over time, while C represents parameters that are calculated as multi-year averages.

3. Results

3.1. Model Validation

The observed NPP and the NPP simulated by the CASA model were used to evaluate the performance of the NPP simulated by the CLM model. The comparison of the measured NPP and the corresponding NPP simulated by CLM 5.0 is shown in Figure 3. The coefficient correlation R^2 is 0.70, which has passed the 95% level with the F test. Figure 4 shows the spatial patterns of the NPP simulated by the CASA model and by CLM 5.0. It has been found that the NPP simulated by CLM 5.0 agrees with the NPP simulated by the CASA model in most areas of the Tibetan Plateau. Compared to the CASA model, CLM 5.0 simulated the spatial patterns of NPP in detail (Figure 4), although there are some regional differences. The NPP of the TP decreases from southeast to northwest, and the highest NPP areas are located on the southern flank of the Eastern Himalayas and in the high mountains of Sichuan and Tibet. Low-NPP areas are in the highland deserts, including the desert areas of the Qaidam Basin, the desert areas of the Kunlun Mountains, and the desert areas of the Ngari Prefecture.

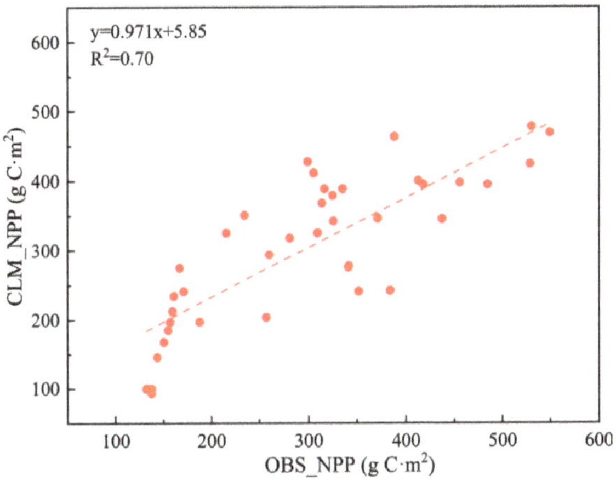

Figure 3. Comparison of the measured NPP and the NPP simulated by CLM 5.0.

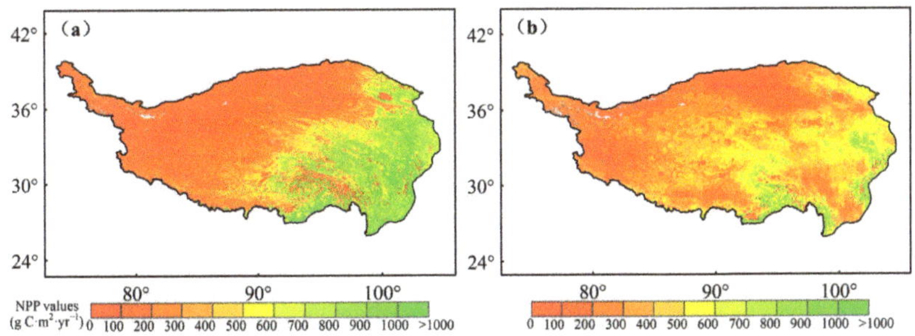

Figure 4. Spatial patterns of NPP simulated by (**a**) CASA model and (**b**) CLM 5.0 from 2000 to 2018.

3.2. Changing Trend in NPP and Driving Factors of the Tibetan Plateau

The NPP simulated by the CLM5.0 model is shown in Figure 5. The multi-year average of NPP simulated by the CLM model for the period 1979–2018 is 256 (g C·m^2·yr^{-1}). The NPP shows an increasing trend over 40 years with an average value of 2.38 (g C·m^2·yr^{-1}).

NPP is affected by different meteorological factors; thus, the NPP in the Tibetan Plateau showed large variations over four decades. Generally, the NPP trend that considers all meteorological factors has increased in most areas of the Tibetan Plateau (Figure 6a). Precipitation led to an increasing trend of the NPP in the central and west–central regions, while the decreasing trend occurred in the central–eastern and southeast regions (Figure 6b). Radiation change led to an increasing trend in the central region (Figure 6c). Temperature change led to an increasing trend in the central and east–central regions of the TP, while the decreasing trend in the western and southern regions of the TP (Figure 6d).

The interannual change of NPP under consideration of different meteorological factors is shown in Figure 7. As shown in Figure 7, NPP exhibited a noteworthy increase under the influence of precipitation, with a rate of 2.4 (g C·m^2·yr^{-1}) ($p < 0.05$). During the period 1979–2018, the interannual variations of NPP under precipitation matched closely with the control experiment, which indicated that precipitation was the main contributor to the interannual variation of NPP. The precipitation accounts for approximately 91% of the NPP variation (Figure 7). Under the influence of temperature, NPP changed at a rate of 0.14 (g C·m^2·yr^{-1}), indicating a slight upward trend, accounting for 5.3% of the interannual variation of NPP. Conversely, the radiation factor led to a decrease in NPP by −0.1 (g C·m^2·yr^{-1}), suggesting a weak declining trend and accounting for 3.8% of the NPP change. However, it is noteworthy that NPP was not significantly affected by temperature and radiation factors.

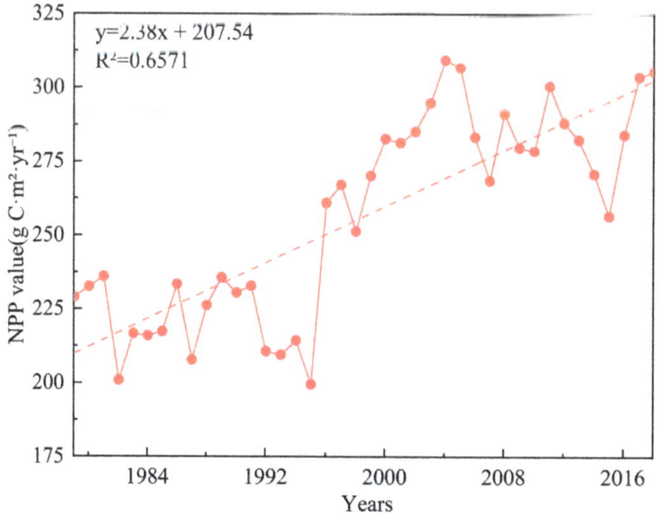

Figure 5. Trends in the simulated NPP by CLM 5.0 from 1979 to 2018.

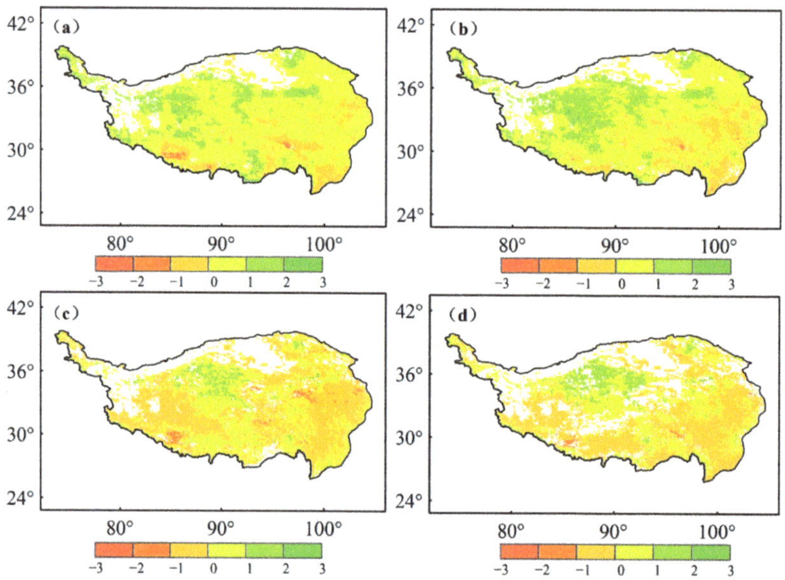

Figure 6. Spatial trends in NPP in the Tibetan Plateau region for the period 1979–2018 under different environmental factors, white areas indicate invalid values. (**a**) All factors. (**b**) Precipitation. (**c**) Radiation. (**d**) Temperature.

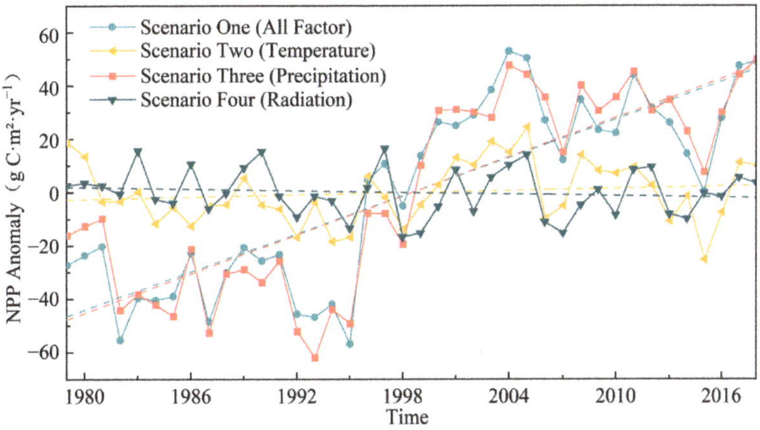

Figure 7. Spatial trends in NPP in the Tibetan Plateau region for the period 1979–2018 under different environmental factors, the dashed lines represents a linear trend.

3.3. Interannual Variation of NPP in Grasslands with Different Coverage and Response to Meteorological Factors

The grasslands are the main land use type in the Tibetan Plateau region. The difference in NPP changes across different grasslands in the TP and their driving meteorological factors were analyzed. The alpine grassland in the TP was divided into high-coverage grasslands, medium-coverage grasslands, and low-coverage grasslands. Although grasslands of all three types are widely distributed over the plateau, their distribution patterns vary significantly (Figure 8a). There is a relatively small area of high-coverage grasslands on the plateau, primarily found in the eastern and southern parts of the plateau, while the plateau's central portion is dominated by grasslands with medium cover. Grasslands with low cover have the greatest distribution range, and they are widely distributed throughout the TP. As shown in Figure 8b, the high-coverage grasslands, the medium-coverage grasslands and the low-coverage grasslands have an average annual NPP of 306, 259 and 243 (g C·m^2·yr^{-1}), respectively. The NPP in low-coverage grasslands, medium-coverage grasslands, and high-coverage grasslands increases significantly by 2.61, 2.39 and 1.97 (g C·m^2·yr^{-1}), respectively.

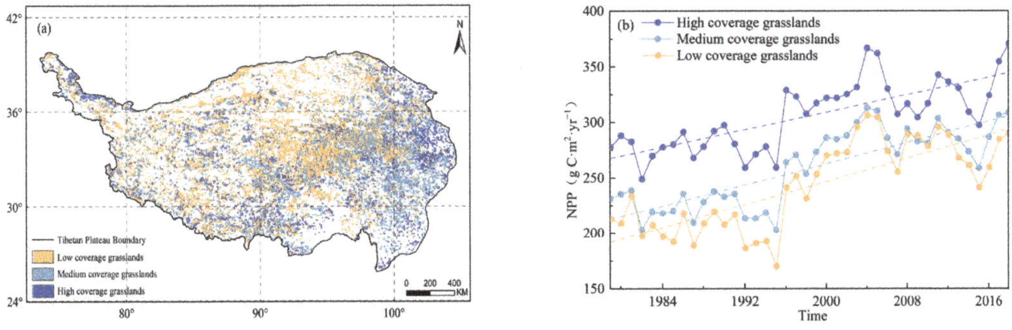

Figure 8. (**a**) Distribution of grasslands with different coverage types in the Tibetan Plateau. (**b**) Temporal Trends in NPP for grasslands with different coverage types, the dashed lines represents a linear trend.

Compared with temperature and radiation, precipitation is the key factor that affects the NPP of all three types grassland (Figure 9). Under the influence of precipitation, the increase rates of high-, medium-, and low-coverage grasslands NPP were 2.0, 2.45, and

2.8 (g C·m²·yr⁻¹), respectively. Low-coverage grassland had the highest rate of increase (Figure 9a). Under the influence of radiation factors, grasslands with high covers increased at the fastest rate of 0.45 (g $C \cdot m^2 \cdot yr^{-1}$) (Figure 9b). In contrast, medium-coverage and low-coverage grasslands have a lower increase rate of 0.09 and 0.02 (g $C \cdot m^2 \cdot yr^{-1}$), respectively. Under the influence of temperature factors, high-coverage grasses show an increasing trend of 0.36 (g $C \cdot m^2 \cdot yr^{-1}$), but the medium- and low-coverage grasses showed a decreasing trend of −0.15 and −0.33 (g $C \cdot m^2 \cdot yr^{-1}$), respectively.

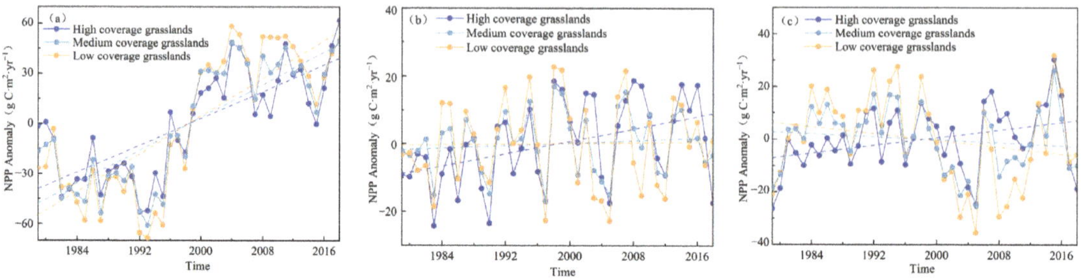

Figure 9. Trends in NPP for high, medium, and low coverage grasslands under different meteorological factors for the period 1979–2018, the dashed lines represents a linear trend. (**a**) Precipitation factor. (**b**) Radiation factor. (**c**) Temperature factor.

4. Discussion

4.1. Impact of Meteorological Factors on NPP Changes

The Tibetan Plateau has undergone significant warming and humidification over the past four decades [56]. Mean annual precipitation and mean annual temperature increased at a rate of 3.45 mm/yr and 0.04 °C/yr, respectively (Figure 10). In contrast, solar radiation displayed a decreasing trend, with a rate of $-0.11\ W \cdot m^2 \cdot yr^{-1}$. Global warming has an impact on vegetation productivity, ecological processes and biodiversity dynamics. The length of the growing season (LOS) grassland becomes longer [57], and the normalized difference vegetation index (NDVI) and mean species abundance (MSA) show increasing trends [58,59].

NPP increased in most areas of the TP except in the eastern and southeastern of the Hengduan Mountains. We found that the precipitation was the main factor leading to the increase in vegetation productivity with the rate of 2.4 (g $C \cdot m^2 \cdot yr^{-1}$) (Figures 6 and 7). This conclusion is consistent with the research from Zeng et al. and Li et al. [60,61]. The vegetation growth is influenced by water and heat conditions [62]. Soil moisture content is directly influenced by precipitation and plays a crucial role in determining plant available water capacity and photosynthetic processes [63]. At the same time, adequate soil moisture ensures nutrient transport and accelerates the mineralization of organic matter in the soil [64,65].

The temporal and spatial trends of NPP on the TP were found to be significantly correlated with precipitation but not significantly correlated with temperature. This result is consistent with previous investigations [66]. The temperature increased significantly on the TP, and higher temperatures can promote enzymatic activities in photosynthesis, thus enhancing the intensity of plant photosynthesis and increasing the absorption of organic matter by plants [67]. However, our study has revealed that the temperature increase did not have a significant impact on the NPP change. Previous studies have discovered that temperature impacts both the rates of photosynthesis and autotrophic respiration (AR). The rising temperatures have a greater impact on the AR than the GPP [68], especially on the TP [69], which explains the slight downward trend of the NPP.

Solar radiation, as the sole energy source for plant photosynthesis, can directly affect vegetation photosynthesis trends. In our study, the slight decreasing trend in solar radiation over the 40-year period resulted in a negligible effect on the NPP, with a rate of −0.10 (g $C \cdot m^2 \cdot yr^{-1}$), suggesting no significant impact (Figure 7).

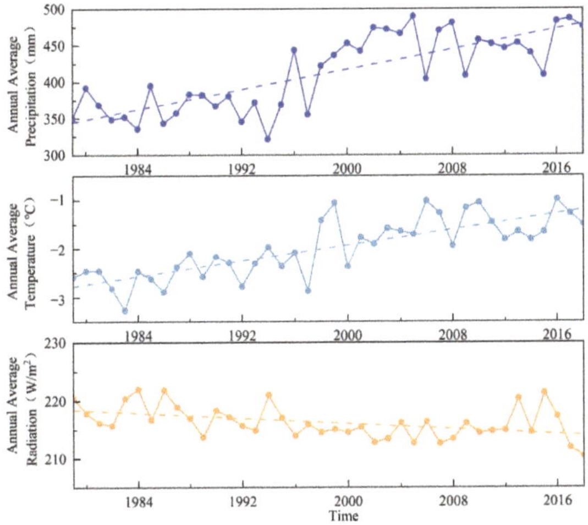

Figure 10. Temporal trends of meteorological factors (precipitation, radiation, temperature) from 1979 to 2018, the dashed lines represents a linear trend.

4.2. Response of NPP to Meteorological Factors in Different Grasslands Cover Types

Grasslands with different cover types displayed different variation characteristics. High-coverage grasslands had a low increase rate, but low-coverage grasslands had a high increase rate. The NPP variation of medium-coverage grasslands and their responses to climate factors fall between high-coverage and low-coverage grasslands. Meteorological factors play a crucial role in shaping these trends (Figure 8b). Our research provides insights into how climate change interacts with all three types of grasslands of the TP.

Precipitation is the most significant factor affecting NPP in low-coverage grasslands, and radiation has no effect, while temperature has a negative effect (Figure 9). This is due to the geographic distribution of low-coverage grasslands in the TP. On the one hand, precipitation increases in low-coverage grasslands, mitigating the impact of soil water deficits. It has been demonstrated that wetter and warmer climatic conditions will alleviate low-temperature constraints and promote the growth of alpine vegetation [70,71]. In regions with lower precipitation, NPP shows a rapid increase with precipitation until it reaches a threshold [72]. On the other hand, low-coverage grassland is predominantly situated in the permafrost region of the west-central plateau, and more permafrost water is released into the soil under the impact of global warming. However, higher temperatures may enhance soil evapotranspiration, exacerbating soil moisture deficits at higher elevations and leading to a declining trend in the effect of temperature on NPP. Previous studies have indicated that climate warming may impede vegetation growth on the TP by increasing soil moisture deficits, which is consistent with the findings of this study [73,74]. Precipitation is the dominant factor of NPP change in high-coverage grassland, followed by radiation and temperature factors. However, the NPP change of high-coverage grassland in the eastern and southern regions of the TP is influenced most by temperature and radiation. That is because seasonally frozen soil maintains relatively high soil moisture content during the growing season and it promotes the impact of radiation and temperature.

At the same time, it is important to consider the influence of non-meteorological factors. China has implemented ecological restoration projects, which have helped reduce grassland degradation caused by grazing. Li et al. [75] found that grazing intensity (14.5%) had a lower impact on controlling Above-ground Net Primary Production (ANPP) grassland dynamics compared to precipitation (29.0%). However, precipitation variability was identified as the key factor for increasing alpine grassland ANPP across the plateau. Meanwhile, under different precipitation patterns, summer precipitation can provide a more effective water supply to the vegetation than meltwater from snowfall or glaciers [69]. The NPP of grasslands on the Tibetan Plateau is expected to increase due to the combined effects of meteorological and non-meteorological factors. In terms of stability, precipitation and temperature variability have a greater impact on the stability of alpine meadows, while temperature variability does not significantly affect grassland stability [76]. It is predicted that the vulnerability of alpine meadows to climate change in the west-central Tibetan Plateau will decrease in the future [77].

5. Conclusions

The study investigates the spatial and temporal trends of NPP and the effects of meteorological factors on the NPP change on the Tibetan Plateau. The annual average NPP is 256 (g $C{\cdot}m^2{\cdot}yr^{-1}$) in the TP. NPP increased at the rate of 2.38 (g $C{\cdot}m^2{\cdot}yr^{-1}$) on the Tibetan Plateau during the period 1979–2018. The variation of the NPP primarily results from precipitation increase, followed by temperature increase and the radiation decrease. Changes in precipitation, temperature and radiation account for approximately 91%, 5.3%, and 3.8% of NPP variation, respectively.

Moreover, our study demonstrates the meteorological drivers of NPP change in the grasslands with different vegetation covers. The annual average NPP of high-coverage grasslands and low-coverage grasslands were 306 and 243 (g $C{\cdot}m^2{\cdot}yr^{-1}$), respectively, and they increased at the rate of 1.97 and 2.61 (g $C{\cdot}m^2{\cdot}yr^{-1}$), respectively. The NPP change of the high-coverage grasslands was mainly affected by precipitation increase (2.0 g $C{\cdot}m^2{\cdot}yr^{-1}$), and then the temperature increase (0.36 g $C{\cdot}m^2{\cdot}yr^{-1}$) and radiation decrease (0.45 g $C{\cdot}m^2{\cdot}yr^{-1}$). Comparatively, the precipitation increase is the driving factor of the increased NPP of low-coverage grasslands (2.82 g $C{\cdot}m^2{\cdot}yr^{-1}$), but the temperature increase is the negative factor (-0.33 g $C{\cdot}m^2{\cdot}yr^{-1}$).

Author Contributions: M.L.: conceptualization, analysis, methodology, data resources and curation, visualization, and writing—original draft. Q.L.: funding acquisition, project administration, supervision, and writing—review and editing. M.X.: software and validation. All authors have read and agreed to the published version of the manuscript.

Funding: This research was funded by Qinghai Provincial Science and Technology Department Major Scientific And Technological Specialties (2021-SF-A6).

Institutional Review Board Statement: Not applicable.

Informed Consent Statement: Not applicable.

Data Availability Statement: The data presented in this study are available on request from the corresponding author. The data are not publicly available due to privacy.

Conflicts of Interest: The authors declare no conflicts of interest.

Appendix A

Table A1. Model data sources.

Data Name	Variables	Data Sources
Data required for the model	Slope, Aspect, Elevation, PFT, etc.	NCAR. (https://svn-ccsm-inputdata.cgd.ucar.edu/trunk/inputdata/) Accessed on 10 December 2023.
Climate forcing dataset	Temperature (K) Pressure (Pa) Specific humidity (Kg/Kg) Wind speed (M/s) Downward shortwave radiation(W/m^2) Downward longwave radiation (W/m^2) Precipitation rate (mm/s)	National Tibetan Plateau/Third Pole Environment Data Center. (http://data.tpdc.ac.cn/en/data/8028b944-daaa-4511-8769-965612652c49/) Accessed on 10 December 2023.
Land use data	High-coverage grasslands Medium-coverage grasslands Low-coverage grasslands	Data Center for Resources and Environmental Sciences (RESDC). (http://www.resdc.cn) Accessed on 20 December 2023.

References

1. Jiang, Y.; Wang, P.; Xu, X.; Zhang, J. Dynamics of carbon fluxes with responses to vegetation, meteorological and terrain factors in the south-eastern Tibetan Plateau. *Environ. Earth Sci.* **2014**, *72*, 4551–4565. [CrossRef]
2. Chen, H.; Zhu, Q.; Peng, C.; Wu, N.; Wang, Y.; Fang, X.; Gao, Y.; Zhu, D.; Yang, G.; Tian, J.; et al. The impacts of climate change and human activities on biogeochemical cycles on the Qinghai-Tibetan Plateau. *Glob. Chang. Biol.* **2013**, *19*, 2940–2955. [CrossRef] [PubMed]
3. Chen, S.; Huang, Y.; Gao, S.; Wang, G. Impact of physiological and phenological change on carbon uptake on the Tibetan Plateau revealed through GPP estimation based on spaceborne solar-induced fluorescence. *Sci. Total Environ.* **2019**, *663*, 45–59. [CrossRef] [PubMed]
4. Field, C.B.; Behrenfeld, M.J.; Randerson, J.T.; Falkowski, P. Primary Production of the Biosphere: Integrating Terrestrial and Oceanic Components. *Science* **1998**, *281*, 237–240. [CrossRef] [PubMed]
5. Luo, Z.; Wu, W.; Yu, X.; Song, Q.; Yang, J.; Wu, J.; Zhang, H. Variation of Net Primary Production and Its Correlation with Climate Change and Anthropogenic Activities over the Tibetan Plateau. *Remote Sens.* **2018**, *10*, 1352. [CrossRef]
6. Liu, C.; Dong, X.; Liu, Y. Changes of NPP and their relationship to climate factors based on the transformation of different scales in Gansu, China. *Catena* **2015**, *125*, 190–199. [CrossRef]
7. Shi, Z.; Wang, Y.; Zhao, Q.; Zhang, L.; Zhu, C. The spatiotemporal changes of NPP and its driving mechanisms in China from 2001 to 2020. *Ecol. Environ. Sci.* **2022**, *31*, 2111–2123.
8. Yang, H.; Zhong, X.; Deng, S.; Xu, H. Assessment of the impact of LUCC on NPP and its influencing factors in the Yangtze River basin, China. *Catena* **2021**, *206*, 105542. [CrossRef]
9. Zhang, M.; Lal, R.; Zhao, Y.; Jiang, W.; Chen, Q. Estimating net primary production of natural grassland and its spatio-temporal distribution in China. *Sci. Total Environ.* **2016**, *553*, 184–195. [CrossRef]
10. Yu, R. An improved estimation of net primary productivity of grassland in the Qinghai-Tibet region using light use efficiency with vegetation photosynthesis model. *Ecol. Model.* **2020**, *431*, 109121. [CrossRef]
11. Zhou, B.; Liao, Z.; Chen, S.; Jia, H.; Zhu, J.; Fei, X. Net Primary Productivity of Forest Ecosystems in the Southwest Karst Region from the Perspective of Carbon Neutralization. *Forests* **2022**, *13*, 1367. [CrossRef]
12. Bayat, M.; Bettinger, P.; Heidari, S.; Hamidi, S.K.; Jaafari, A. A Combination of Biotic and Abiotic Factors and Diversity Determine Productivity in Natural Deciduous Forests. *Forests* **2021**, *12*, 1450. [CrossRef]
13. Bayat, M.; Burkhart, H.; Namiranian, M.; Hamidi, S.K.; Heidari, S.; Hassani, M. Assessing Biotic and Abiotic Effects on Biodiversity Index Using Machine Learning. *Forests* **2021**, *12*, 461. [CrossRef]
14. Prăvălie, R.; Niculiță, M.; Roșca, B.; Marin, G.; Dumitrașcu, M.; Patriche, C.; Birsan, M.; Nita, I.; Tișcovschi, A.; Sîrodoev, I.; et al. Machine learning-based prediction and assessment of recent dynamics of forest net primary productivity in Romania. *J. Environ. Manag.* **2023**, *334*, 117513. [CrossRef] [PubMed]
15. Bulut, S.; Günlü, A.; Şatır, O. Estimating net primary productivity of semi-arid Crimean pine stands using biogeochemical modelling, remote sensing, and machine learning. *Ecol. Inform.* **2023**, *76*, 102137. [CrossRef]
16. Yan, H.; Ran, Q.; Hu, R.; Xue, K.; Zhang, B.; Zhou, S.; Zhang, Z.; Tang, L.; Che, R.; Pang, Z.; et al. Machine learning-based prediction for grassland degradation using geographic, meteorological, plant and microbial data. *Ecol. Indic.* **2022**, *137*, 108738. [CrossRef]
17. Lou, P.; Wu, T.; Yang, S.; Wu, X.; Chen, J.; Zhu, X.; Chen, J.; Lin, X.; Li, R.; Shang, C.; et al. Deep learning reveals rapid vegetation greening in changing climate from 1988 to 2018 on the Qinghai-Tibet Plateau. *Ecol. Indic.* **2023**, *148*, 110020. [CrossRef]
18. Friedman, J. Greedy Function Approximation: A Gradient Boosting Machine. *Ann. Stat.* **2000**, *29*, 1189–1232. [CrossRef]
19. Chen, T.; Guestrin, C. Xgboost: A scalable tree boosting system. In Proceedings of the 22nd Acm Sigkdd International Conference on Knowledge Discovery and Data Mining, San Francisco, CA, USA, 13–17 August 2016; pp. 785–794.
20. Ke, G.; Meng, Q.; Finley, T.; Wang, T.; Chen, W.; Ma, W.; Ye, Q.; Liu, T. LightGBM: A Highly Efficient Gradient Boosting Decision Tree. *Adv. Neural Inf. Process. Syst.* **2017**, *30*, 3149–3157.
21. Wang, T.; Lin, X.; Liu, Y.; Dantec-Nédélec, S.; Ottle, C. Causes of uncertainty in China's net primary production over the 21st century projected by the CMIP5 Earth system models. *Int. J. Climatol.* **2016**, *36*, 2323–2334. [CrossRef]
22. Zhang, X.; Peng, S.; Ciais, P.; Wang, Y.; Silver, J.D.; Piao, S.; Rayner, P.J. Greenhouse Gas Concentration and Volcanic Eruptions Controlled the Variability of Terrestrial Carbon Uptake Over the Last Millennium. *J. Adv. Model. Earth Syst.* **2019**, *11*, 1715–1734. [CrossRef] [PubMed]
23. Ren, H.; Zhang, L.; Yan, M.; Tian, X.; Zheng, X. Sensitivity analysis of Biome-BGCMuSo for gross and net primary productivity of typical forests in China. *For. Ecosyst.* **2022**, *9*, 100011. [CrossRef]
24. Lu, Y.; Huang, Y.; Zhuang, Q.; Sun, W.; Chen, S.; Lu, J. China's Terrestrial Ecosystem Carbon Balance During the 20th Century: An Analysis with a Process-Based Biogeochemistry Model. *Carbon Balanc. Manag.* **2022**, *17*, 16. [CrossRef] [PubMed]
25. Bonan, G.; Lombardozzi, D.; Wieder, W.; Oleson, K.; Lawrence, D.; Hoffman, F.; Collier, N. Model Structure and Climate Data Uncertainty in Historical Simulations of the Terrestrial Carbon Cycle (1850–2014). *Glob. Biogeochem. Cycle* **2019**, *33*, 1310–1326. [CrossRef]
26. Ferro Duarte, H.; Raczka, B.; Ricciuto, D.; Lin, J.; Koven, C.; Thornton, P.; Bowling, D.; Lai, C.; Bible, K.; Ehleringer, J. Evaluating the Community Land Model (CLM4.5) at a coniferous forest site in northwestern United States using flux and carbon-isotope measurements. *Biogeosciences* **2017**, *14*, 4315–4340. [CrossRef]

27. Lin, S.; Wang, G.; Hu, Z.; Huang, K.; Sun, J.; Sun, X. Spatiotemporal Variability and Driving Factors of Tibetan Plateau Water Use Efficiency. *J. Geophys. Res. Atmos.* **2020**, *125*, e2020JD032642. [CrossRef]
28. Hao, R.; Yu, D.; Huang, T.; Li, S.; Qiao, J. NPP plays a constraining role on water-related ecosystem services in an alpine ecosystem of Qinghai, China. *Ecol. Indic.* **2022**, *138*, 108846. [CrossRef]
29. Reich, P.B.; Hobbie, S.E.; Lee, T.D.; Rich, R.; Pastore, M.A.; Worm, K. Synergistic effects of four climate change drivers on terrestrial carbon cycling. *Nat. Geosci.* **2020**, *13*, 787–793. [CrossRef]
30. Tu, H.; Jiapaer, G.; Yu, T.; Li, X.; Chen, B. Analysis of spatio-temporal variation characteristics and influencing factors of net primary productivity in terrestrial ecosystems of China. *Acta Ecol. Sin.* **2023**, *43*, 1219–1233.
31. Zha, X.; Niu, B.; Li, M.; Duan, C. Increasing Impact of Precipitation on Alpine-Grassland Productivity over Last Two Decades on the Tibetan Plateau. *Remote Sens.* **2022**, *14*, 3430. [CrossRef]
32. Zheng, Z.; Zhu, W.; Zhang, Y. Seasonally and spatially varied controls of climatic factors on net primary productivity in alpine grasslands on the Tibetan Plateau. *Glob. Ecol. Conserv.* **2020**, *21*, e00814. [CrossRef]
33. Qiu, L.; Liu, X. Sensitivity analysis of modelled responses of vegetation dynamics on the Tibetan Plateau to doubled CO_2 and associated climate change. *Theor. Appl. Climatol.* **2016**, *124*, 229–239. [CrossRef]
34. Lu, Y.; Yang, X. Using the anomaly forcing Community Land Model (CLM 4.5) for crop yield projections. *Geosci. Model Dev.* **2021**, *14*, 1253–1265. [CrossRef]
35. Hurrell, J.; Holland, M.; Gent, P.; Ghan, S.; Kay, J.; Kushner, P.; Lamarque, J.; Large, W.; Lawrence, D.; Marshall, S.; et al. The community earth system model: A framework for collaborative research. *Bull. Am. Meteorol. Soc.* **2013**, *94*, 1339–1360. [CrossRef]
36. Lawrence, D.M.; Fisher, R.A.; Koven, C.D.; Oleson, K.W.; Swenson, S.C.; Bonan, G.; Collier, N.; Ghimire, B.; van Kampenhout, L.; Kennedy, D.; et al. The Community Land Model Version 5: Description of New Features, Benchmarking, and Impact of Forcing Uncertainty. *J. Adv. Model. Earth Syst.* **2019**, *11*, 4245–4287. [CrossRef]
37. Schädel, C.; Rogers, B.M.; Lawrence, D.M.; Koven, C.D.; Brovkin, V.; Burke, E.J.; Genet, H.; Huntzinger, D.N.; Jafarov, E.; McGuire, A.D.; et al. Earth system models must include permafrost carbon processes. *Nat. Clim. Chang.* **2024**, *14*, 114–116. [CrossRef]
38. Raczka, B.; Hoar, T.J.; Duarte, H.F.; Fox, A.M.; Anderson, J.L.; Bowling, D.R.; Lin, J.C. Improving CLM5.0 Biomass and Carbon Exchange Across the Western United States Using a Data Assimilation System. *J. Adv. Model. Earth Syst.* **2021**, *13*, e2020MS002421. [CrossRef] [PubMed]
39. Zhang, L.; Ning, F.; Bai, X.; Zeng, X.; He, C. Performance evaluation of CLM5.0 in simulating liquid soil water in high mountainous area, Northwest China. *J. Mt. Sci.* **2023**, *20*, 1865–1883. [CrossRef]
40. Luo, Q.; Wen, J.; Hu, Z.; Lu, Y.; Yang, X. Parameter Sensitivities of the Community Land Model at Two Alpine Sites in the Three-River Source Region. *J. Meteorol. Res.* **2020**, *34*, 851–864. [CrossRef]
41. Yang, Q.; Zhu, X.; Ou, G.; Ma, L. Estimation of CLM5.0 Parameters for Improving Grassland Productivity Simulation in Hulunburi, Inner Mongolia. In Proceedings of the IGARSS 2023—2023 IEEE International Geoscience and Remote Sensing Symposium, Pasadena, CA, USA, 16–21 July 2023; pp. 3307–3310.
42. Ma, X.; Wang, A. Systematic Evaluation of a High-Resolution CLM5 Simulation over Continental China for 1979–2018. *J. Hydrometeorol.* **2022**, *23*, 1879–1897. [CrossRef]
43. Yang, S.; Li, R.; Zhao, L.; Wu, T.; Wu, X.; Zhang, Y.; Shi, J.; Qiao, Y. Evaluation of the Performance of CLM5.0 in Soil Hydrothermal Dynamics in Permafrost Regions on the Qinghai–Tibet Plateau. *Remote Sens.* **2022**, *14*, 6228. [CrossRef]
44. Fang, Y.; Liu, C.; Leung, L.R. Accelerating the spin-up of the coupled carbon and nitrogen cycle model in CLM4. *Geosci. Model Dev.* **2015**, *8*, 781–789. [CrossRef]
45. Li, F.; Zeng, X.; Song, X.; Tian, D.; Shao, P.; Zhang, D. Impact of spin-up forcing on vegetation states simulated by a dynamic global vegetation model coupled with a land surface model. *Adv. Atmos. Sci.* **2011**, *28*, 775–788. [CrossRef]
46. He, J.; Yang, K.; Tang, W.; Lu, H.; Qin, J.; Chen, Y.; Li, X. The first high-resolution meteorological forcing dataset for land process studies over China. *Sci. Data* **2020**, *7*, 25. [CrossRef] [PubMed]
47. Rienecker, M.M.; Suarez, M.J.; Gelaro, R.; Todling, R.; Bacmeister, J.; Liu, E.; Bosilovich, M.G.; Schubert, S.D.; Takacs, L.; Kim, G.; et al. MERRA: NASA's Modern-Era Retrospective Analysis for Research and Applications. *J. Clim.* **2011**, *24*, 3624–3648. [CrossRef]
48. Pinker, R.T.; Tarpley, J.D.; Laszlo, I.; Mitchell, K.E.; Houser, P.R.; Wood, E.F.; Schaake, J.C.; Robock, A.; Lohmann, D.; Cosgrove, B.A.; et al. Surface radiation budgets in support of the GEWEX Continental-Scale International Project (GCIP) and the GEWEX Americas Prediction Project (GAPP), including the North American Land Data Assimilation System (NLDAS) project. *J. Geophys. Res. Atmos.* **2003**, *108*, D22. [CrossRef]
49. Yang, F.; Lu, H.; Yang, K.; He, J.; Wang, W.; Wright, J.S.; Li, C.; Han, M.; Li, Y. Evaluation of multiple forcing data sets for precipitation and shortwave radiation over major land areas of China. *Hydrol. Earth Syst. Sci.* **2017**, *21*, 5805–5821. [CrossRef]
50. Xu, X.; Liu, J.; Zhang, S.; Li, R.; Yan, C.; Wu, S. *China Multi-Period Land Use and Land Cover Remote Sensing Monitoring Dataset (CNLUCC)*; Data Registration and Publishing System of Resource and Environmental Science Data Center, Chinese Academy of Sciences: Beijing, China, 2018.
51. Zeng, J.; Zhou, T.; Xu, Y.; Lin, Q.; Tan, E.; Zhang, Y.; Wu, X.; Zhang, J.; Liu, X. The fusion of multiple scale data indicates that the carbon sink function of the Qinghai-Tibet Plateau is substantial. *Carbon Balanc. Manag.* **2023**, *18*, 19. [CrossRef]
52. Liu, Y.; Zhou, R.; Ren, H.; Zhang, W.; Zhang, Z.; Zhang, Z.; Wen, Z. Evaluating the dynamics of grassland net primary productivity in response to climate change in China. *Glob. Ecol. Conserv.* **2021**, *28*, e01574. [CrossRef]

53. Xu, H.; Wang, X.; Zhang, X. Alpine grasslands response to climatic factors and anthropogenic activities on the Tibetan Plateau from 2000 to 2012. *Ecol. Eng.* **2016**, *92*, 251–259. [CrossRef]
54. Wang, S.; Zhang, B.; Yang, Q.; Chen, G.; Yang, B.; Lu, L.; Shen, M.; Peng, Y. Responses of net primary productivity to phenological dynamics in the Tibetan Plateau, China. *Agric. For. Meteorol.* **2017**, *232*, 235–246. [CrossRef]
55. Zhou, W.; Wang, T.; Xiao, J.; Wang, K.; Yu, W.; Du, Z.; Huang, L.; Yue, T. Grassland productivity increase was dominated by climate in Qinghai-Tibet Plateau from 1982 to 2020. *J. Clean Prod.* **2024**, *434*, 140144. [CrossRef]
56. Li, L.; Li, H.; Shen, H.; Liu, C.; Ma, Y.; Zhao, Y. The truth and inter-annual oscillation causes for climate change in the Qinghai-Tibet Plateau. *J. Glaciol. Geocryol.* **2018**, *40*, 1079–1089.
57. Zhong, R.; Yan, K.; Gao, S.; Yang, K.; Zhao, S.; Ma, X.; Zhu, P.; Fan, L.; Yin, G. Response of grassland growing season length to extreme climatic events on the Qinghai-Tibetan Plateau. *Sci. Total Environ.* **2024**, *909*, 168488. [CrossRef]
58. Pang, G.; Wang, X.; Yang, M. Using the NDVI to identify variations in, and responses of, vegetation to climate change on the Tibetan Plateau from 1982 to 2012. *Quat. Int.* **2017**, *444*, 87–96. [CrossRef]
59. Liu, H.; Liu, S.; Wang, F.; Liu, Y.; Han, Z.; Wang, Q.; Yu, L.; Dong, Y.; Sun, J. Multilevel driving factors affecting ecosystem services and biodiversity dynamics on the Qinghai-Tibet Plateau. *J. Clean. Prod.* **2023**, *396*, 136448. [CrossRef]
60. Zeng, N.; Ren, X.; He, H.; Zhang, L.; Niu, Z. Precipitation Conditions Constrain the Sensitivity of Aboveground Net Primary Productivity in Tibetan Plateau Grasslands to Climate Change. *Remote Sens.* **2023**, *15*, 2591. [CrossRef]
61. Li, H.; Wu, Y.; Liu, S.; Xiao, J. Regional contributions to interannual variability of net primary production and climatic attributions. *Agric. For. Meteorol.* **2021**, *303*, 108384. [CrossRef]
62. Ye, C.; Sun, J.; Liu, M.; Xiong, J.; Zong, N.; Hu, J.; Huang, Y.; Duan, X.; Tsunekawa, A. Concurrent and Lagged Effects of Extreme Drought Induce Net Reduction in Vegetation Carbon Uptake on Tibetan Plateau. *Remote Sens.* **2020**, *12*, 2347. [CrossRef]
63. Gang, C.; Zhou, W.; Chen, Y.; Zhaoqi, W.; Sun, Z.; Li, J.; Qi, J.; Odeh, I. Quantitative assessment of the contributions of climate change and human activities on global grassland degradation. *Environ. Earth Sci.* **2014**, *72*, 4273–4282. [CrossRef]
64. Wang, J.; Sun, H.; Xiong, J.; He, D.; Cheng, W.; Ye, C.; Yong, Z.; Huang, X. Dynamics and Drivers of Vegetation Phenology in Three-River Headwaters Region Based on the Google Earth Engine. *Remote Sens.* **2021**, *13*, 2528. [CrossRef]
65. Yang, X.; Wu, J.; Chen, X.; Ciais, P.; Maignan, F.; Yuan, W.; Piao, S.; Yang, S.; Gong, F.; Su, Y.; et al. A comprehensive framework for seasonal controls of leaf abscission and productivity in evergreen broadleaved tropical and subtropical forests. *Innovation* **2021**, *2*, 100154. [CrossRef] [PubMed]
66. Liu, J.; Ji, Y.; Zhou, G.; Zhou, L.; Lv, X.; Zhou, M. Temporal and spatial variations of net primary productivity (NPP) and its climate driving effect in the Qinghai-Tibet Plateau, China from 2000 to 2020. *Chin. J. Appl. Ecol.* **2022**, *33*, 1533–1538.
67. Davidson, E.A.; Janssens, I.A. Temperature sensitivity of soil carbon decomposition and feedbacks to climate change. *Nature* **2006**, *440*, 165–173. [CrossRef] [PubMed]
68. He, Y.; Piao, S.; Li, X.; Chen, A.; Qin, D. Global patterns of vegetation carbon use efficiency and their climate drivers deduced from MODIS satellite data and process-based models. *Agric. For. Meteorol.* **2018**, *256–257*, 150–158. [CrossRef]
69. Lin, S.; Wang, G.; Feng, J.; Dan, L.; Sun, X.; Hu, Z.; Chen, X.; Xiao, X. A carbon flux assessment driven by environmental factors over the Tibetan Plateau and various permafrost regions. *J. Geophys. Res. Biogeosci.* **2019**, *124*, 1132–1147. [CrossRef]
70. Zhang, G.; Zhang, Y.; Dong, J.; Xiao, X. Green-up dates in the Tibetan Plateau have continuously advanced from 1982 to 2011. *Proc. Natl. Acad. Sci. USA* **2013**, *110*, 4309–4314. [CrossRef]
71. Shen, M.; Wang, S.; Jiang, N.; Sun, J.; Cao, R.; Ling, X.; Fang, B.; Zhang, L.; Zhang, L.; Xu, X.; et al. Publisher Correction: Plant phenology changes and drivers on the Qinghai–Tibetan Plateau. *Nat. Rev. Earth Environ.* **2022**, *3*, 717. [CrossRef]
72. Sun, J.; Du, W. Effects of precipitation and temperature on net primary productivity and precipitation use efficiency across China's grasslands. *GIScience Remote. Sens.* **2017**, *54*, 881–897. [CrossRef]
73. Wang, Y.; Liu, X.; Lv, M.; Zhang, Z. Mechanisms and influencing factors of hydrothermal processes in active layer soils on the Qinghai-Tibet Plateau under freeze–thaw action. *Catena* **2023**, *220*, 106694. [CrossRef]
74. Jin, H.; He, R.; Cheng, G.; Wu, Q.; Wang, S.; Lü, L.; Chang, X. Changes in frozen ground in the Source Area of the Yellow River on the Qinghai–Tibet Plateau, China, and their eco-environmental impacts. *Environ. Res. Lett.* **2009**, *4*, 45206. [CrossRef]
75. Meng, L.; Wu, J.; Feng, Y.; Niu, B.; He, Y.; Zhang, X. Climate Variability Rather Than Livestock Grazing Dominates Changes in Alpine Grassland Productivity Across Tibet. *Front. Ecol. Evol.* **2021**, *9*, 631024.
76. Ren, L.; Huo, J.; Xiang, X.; Pan, Y.; Li, Y.; Wang, Y.; Meng, D.; Yu, C.; Chen, Y.; Xu, Z.; et al. Environmental conditions are the dominant factor influencing stability of terrestrial ecosystems on the Tibetan plateau. *Commun. Earth Environ.* **2023**, *4*, 196. [CrossRef]
77. Yuan, Q.; Wu, S.; Dai, E.; Zhao, D.; Ren, P.; Zhang, X. NPP vulnerability of the potential vegetation of China to climate change in the past and future. *J. Geogr. Sci.* **2017**, *27*, 131–142. [CrossRef]

atmosphere

MDPI

Article

Contribution of Atmospheric Factors in Predicting Sea Surface Temperature in the East China Sea Using the Random Forest and SA-ConvLSTM Model

Qiyan Ji [1], Xiaoyan Jia [1,*], Lifang Jiang [2], Minghong Xie [1], Ziyin Meng [1], Yuting Wang [1] and Xiayan Lin [1]

[1] Marine Science and Technology College, Zhejiang Ocean University, Zhoushan 316022, China; jiqiyan@zjou.edu.cn (Q.J.); xieminghong@zjou.edu.cn (M.X.); mengziyin@zjou.edu.cn (Z.M.); wangyuting@zjou.edu.cn (Y.W.); linxiayan@zjou.edu.cn (X.L.)

[2] South China Sea Marine Forecast and Hazard Mitigation Center, Ministry of Natural Resources, Guangzhou 510310, China; jlf@hyyb.org

* Correspondence: s20070700001@zjou.edu.cn

Abstract: Atmospheric forcings are significant physical factors that influence the variation of sea surface temperature (SST) and are often used as essential input variables for ocean numerical models. However, their contribution to the prediction of SST based on machine-learning methods still needs to be tested. This study presents a prediction model for SST in the East China Sea (ECS) using two machine-learning methods: Random Forest and SA-ConvLSTM algorithms. According to the Random Forest feature importance scores and correlation coefficients R, 2 m air temperature and longwave radiation were selected as the two most important key atmospheric factors that can affect the SST prediction performance of machine-learning methods. Four datasets were constructed as input to SA-ConvLSTM: SST-only, SST-T2m, SST-LWR, and SST-T2m-LWR. Using the SST-T2m and SST-LWR, the prediction skill of the model can be improved by about 9.9% and 9.43% for the RMSE and by about 8.97% and 8.21% for the MAE, respectively. Using the SST-T2m-LWR dataset, the model's prediction skill can be improved by 10.75% for RMSE and 9.06% for MAE. The SA-ConvLSTM can represent the SST in ECS well, but with the highest RMSE and AE in summer. The findings of the presented study requires much more exploration in future studies.

Keywords: Random Forest; SA-ConvLSTM; East China Sea; sea surface temperature; prediction

1. Introduction

The ocean covers about 75% of the Earth's surface, and its thermal energy plays a crucial role in regulating the global climate and ecosystems. Sea surface temperature (SST) is one of the most important ocean parameters with significant impacts on global climate and marine ecosystems [1–4]. Accurate and effective forecasting of SST is important for marine weather services, climate prediction, marine fisheries production, and marine environmental protection [5–7]. The East China Sea (ECS) is a shelf sea in the northwestern Pacific Ocean, located to the west of mainland China, south of the Yellow Sea, and north of the South China Sea [8]. The SST of the ECS is influenced by many factors such as solar radiation, monsoon, East China Sea shelf circulation, Kuroshio intrusion, runoff, tides, etc. [9]. These factors are highly stochastic and uncertain, and pose many challenges to predicting the SST accurately. Researchers and scientists have proposed many methods for predicting SST, which can be divided into two categories: one is based on numerical ocean models, and the other is based on data-driven methods.

The numerical ocean model is a powerful tool for marine science research and operational oceanography, and it is commonly used to simulate the physical, chemical, and biological processes of the ocean [10]. Modern numerical ocean models are mainly based on complex physical equations, which are difficult to solve without certain approximations and simplifications. SST prediction by numerical ocean models usually requires precise initial and boundary conditions and atmospheric forcing [11]. Many typical numerical ocean models, such as HYCOM, ROMS, POM etc., have been widely used to simulate and predict not only the SST but also the three-dimensional temperature field of the ocean [12]. Gao et al. used HYCOM to simulate the SST of the Tropical and North Pacific based on the different combinations of two air–sea flux data sets (COADS and ECMWF) and two bulk parameter formulas (non-constant and constant) [13]. Zhang et al. used the ROMS to simulate the three-dimensional sea temperature in the northern Yellow Sea and analyzed its distribution characteristics and variations [14]. Based on the POM, Zhao et al. constructed a three-dimensional baroclinic assimilation model to model to simulate the current and temperature field in the Bohai Sea, Yellow Sea, and East China Sea [15]. However, strong nonlinear instability, huge computational expense, low reusability efficiency, and high coupling costs have gradually become the main bottlenecks for the further development of numerical ocean modeling. The parameterization schemes of physical and dynamical processes in numerical ocean models, inaccurate initial and boundary conditions, and atmospheric forcing can easily lead to numerical model prediction errors and affect simulation and prediction accuracy.

In recent years, data-driven methods have also been widely used to build SST prediction models by learning the data variability rule from historical SST observations [16]. Data-driven methods rely primarily on observational data rather than solving complex physical governing equations. They are simpler, less computationally intensive, and more likely to achieve higher predictive performance than numerical models. Data-driven methods include traditional statistical models, shallow neural network models, and deep learning models [17–20]. Traditional statistical models such as Markov and regression models are first employed to predict SST. Xue et al. constructed a seasonally varying Markov model to predict SST and sea level in the tropical Pacific [21]. Laepple et al. used a regression model to predict SST and found a correlation between SST and the number of hurricanes [22]. Kug et al. developed a regression model for the dynamic prediction of monthly SST in the Indian Ocean based on the lagged relationship between SST and Nino 3 [23]. The above-mentioned statistical models can predict the evolution of the SST based on specific observational data, but it is difficult to further improve the accuracy of the forecasts due to data and methodological limitations [24]

As increasing SST data have become available, researchers have been trying to use new data-driven methods to predict SST. Shallow neural network models have become increasingly popular due to their high degree of flexibility in fitting SST data [25,26] and have shown substantial improvements in prediction accuracy compared to traditional statistical models. Tangang et al. used a neural network to predict SST anomalies in the equatorial Pacific Niño region, achieving a correlation skill of 0.8, demonstrating the feasibility of neural networks to capture nonlinear relationships [27]. Wu et al. developed a nonlinear prediction system for tropical Pacific SST anomalies using a multilayer neural network approach, which improved the correlation coefficient by 0.1–0.14 compared to a linear regression model [28]. Tripathi et al. used an artificial neural network to predict SST anomalies in a specific region of the Indian Ocean with high accuracy [29]. Aparna et al. (2018) proposed a three-layer neural network to predict SST at a given location with a prediction error of ±0.5 °C [30]. Gupta et al. evaluated the prediction ability of different methods based on specific training, regression, and artificial neural networks and found that the neural networks showed a better RMSE value of 1.3 °C than the other methods [31]. However, many shallow neural networks cannot fully utilize the large amount of historical SST data to train prediction models due to their limited learning capacity and relatively simple structure [32].

In recent years, deep learning models have gained popularity in predicting SST due to their powerful learning and modeling capabilities [9]. Zhang et al. used a Long Short-Term Memory (LSTM) model to forecast SST in the Bohai Sea [33]. They inputted daily, weekly, and monthly SST data and achieved an RMSE value of less than 1.13 °C. Xiao et al.

developed an LSTM-AdaBoost model to predict the SST in the East China Sea region for the next 10 days, and the model achieved a high correlation coefficient of 95% [34]. He et al. proposed an SST prediction model based on the EMD-GRU method and found that the EMD-GRU model reduced mean square error (MSE) and mean absolute error (MAE) to different degrees compared with LSTM and GRU models [35]. Zheng et al. proposed a deep learning model for predicting SST in the equatorial eastern Pacific Ocean, which includes a convolutional layer, a max-pooling layer, and an up-sampling layer [20]. The experimental results show an RMSE value of less than 0.9 °C. Ham et al. proposed a model based on CNN to predict the Nino 3.4 index for the next 23 months and found that the CNN model could predict ENSO 1.5 years in advance with a correlation coefficient of up to 90% [36]. However, these methods only consider the temporal evolution of the SST data without taking into account its spatial correlation.

To address the above problem, Shi et al. transformed the precipitation proximity prediction into a space–time series prediction problem and proposed the ConvLSTM model, which was subsequently used in SST prediction [37]. Xiao et al. used the ConvLSTM model to predict the SST in the East China Sea for the next 10 days, and the results showed that the proposed model outperformed the SVR model and the LSTM mode [16]. Zhang et al. proposed the M-ConvLSTM model for three-dimensional temperature prediction, which showed strong prediction performance with a coefficient of determination R^2 greater than 0.84 at different water depths [38]. Although the ConvLSTM model is effective in predicting SST, it is limited in achieving long-range spatial dependence by stacking convolutional layers because the effective receptive field is much smaller than the theoretical receptive field [39]. To address the above issues, Lin et al. proposed a self-attentive memory module (SAM) for ConvLSTM, which is embedded in the ConvLSTM model to enable it to effectively capture long-range spatial dependencies, referred to as SA-ConvLSTM [40]. Therefore, this paper will use the SA-ConvLSTM model to predict the SST in the East China Sea.

Since most current data-driven approaches ignore physical mechanisms, both the methodology and data are crucial for any data-driven SST prediction model. Although there has been much research on innovative methodologies, further investigation is required to determine the impact of the dataset on prediction accuracy. In addition to the essential SST data, are there other datasets that have a physical relationship with SST that could affect SST forecast accuracy when using a data-driven method? Atmospheric forcings have a strong physical relationship with SST variability and are often considered an essential input for predicting SST using numerical oceanic models. It is worth considering how these data can be used to improve the accuracy of the SST forecasts using the data-driven method that has been selected, namely the SA-ConvLSTM model.

This paper aims to investigate the influence of atmospheric factors on the accuracy of SST predictions using SA-ConvLSTM. The study considers various atmospheric factors, including the 2 m air temperature, sea surface solar shortwave radiation flux, sea surface longwave radiation flux, sea surface latent heat flux, sea surface sensible heat flux, and the net heat flux, wind of 10 m. These factors are considered to be the most important atmospheric factors that can affect the sea surface temperature. The feature importance score of the Random Forest model was used to assess the contribution of these atmospheric factors and select the critical factors to construct different datasets. The SA-ConvLSTM model is used to predict the SST in the East China Sea, and the impact of different datasets on the prediction results of the SA-ConvLSTM model will be analyzed. The remainder of the paper is organized as follows. Section 2 focuses on the data sources and pre-processing. Section 3 provides a detailed account of the methodologies employed in this study and the construction of the model. Section 4 presents the experimental results. Section 5 offers a discussion of the findings. Section 6 presents the conclusion.

2. Data

The high-resolution satellite remotely sensed SST data used in this paper is the Operational Sea Surface Temperature and Ice Analysis (OSTIA) [41]. OSTIA is a daily global SST product developed by the UK Met Office with a resolution of 1/20°. These data are derived from AATSR, SEVIRI, AVHRR, AMSR, and TMI measurements with a root mean square error of less than 0.6 °C. The data can be found on the following website: https://www.ncei.noaa.gov/data/oceans/ghrsst/L4/GLOB/UKMO/OSTIA/ (accessed on 11 November 2021). The paper uses data with a spatial range of 22° N~33° N, 120° E~131° E, and a temporal range of 2010 to 2020. The training data set consists of data from 2010 to 2019, while the test data set consists of data from 2020.

The atmospheric factors data used in this paper include 2 m air temperature (T2m), sea surface solar shortwave radiation flux (*SWR*), sea surface longwave radiation flux (*LWR*), sea surface latent heat flux (LHF), sea surface sensible heat flux (*SHF*), U-component of 10 m wind and V-component of 10 m wind (10 m-wind) and were provided by ERA5. ERA5 is the fifth generation of the European Centre for Medium-Range Weather Forecasts (ECMWF) atmospheric reanalysis of global climate data. ERA5 combines model data with observations to provide a numerical interpretation of the recent climate. It offers hourly data on atmospheric, terrestrial, and oceanic climate variables at a resolution of 1/4°. The data can be found here: https://cds.climate.copernicus.eu/ (accessed on 20 December 2021). The paper uses data with a spatial range of 22° N~33° N, 120° E~131° E, and a temporal range from 2010 to 2020. The training dataset consists of data from 2010 to 2019, while the test dataset consists of data from 2020. The sum of heat flux into or out of the water, which is referred to as the net heat flux *Qnet*, is also considered to be a factor that can have an impact on sea surface temperature. Since we only consider the influence of atmospheric factors, the heat flux carried by ocean currents to the regional sea is not considered in this paper. Refer to the heat budget equation in [42]; the *Qnet* is calculated with the following equation:

$$Qnet = SWR + LWR + LHR + SHF \tag{1}$$

3. Method

3.1. Random Forest

Random Forest (RF) is a non-parametric supervised learning method for classification and regression. It was first proposed by Breiman and can be regarded as a special type of bagging algorithm [43,44]. Random Forest is widely used as a classifier in remote sensing. It can be successfully used to select and rank the variables with the greatest ability to discriminate between the target classes [45]. The Random Forest is also used in various other research projects, including the environmental impact assessment of multi-source solid waste [46], the prediction of perennial ryegrass biomass [47], the identification and control of land subsidence and uplift [48], and the predicting of SST [49,50]. As Random Forest can be used to select and rank the key variables, it is well suited to selecting and ranking the key atmospheric factors that can improve the accuracy of the SST prediction. Random Forest consists of multiple decision trees, each of which is trained on independent data, and a final prediction is obtained by voting or averaging. The algorithm flow is shown in Figure 1. In this paper, we use Random Forests to evaluate the importance of the features that can affect the SST, and then we select the most important features based on their score in the feature importance evaluation.

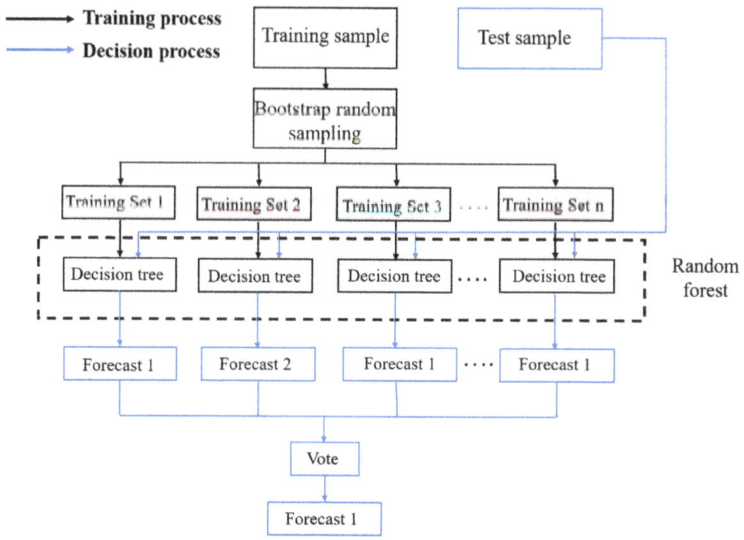

Figure 1. Flow chart of the Random Forest algorithm.

3.2. SA-ConvLSTM Model

The SA-ConvLSTM model was proposed by Lin et al., which can capture the long-range spatial dependence of the data well [40]. The SA-ConvLSTM model is a combination of the self-attention module and ConvLSTM, as shown in Equation (2). However, Lin et al. discovered that this basic model did not yield satisfactory experimental results, prompting them to make improvements to the self-attention module [40].

$$\begin{cases} X_t = SA(\overline{X}_t) \\ \hat{H}_{t-1} = SA(\hat{h}_{t-1}) \\ i_t = \sigma(W_i * [\hat{H}_{t-1}, X_t] + b_i) \\ f_t = \sigma(W_f * [\hat{H}_{t-1}, X_t] + b_f) \\ o_t = \sigma(W_o * [\hat{H}_{t-1}, X_t] + b_o) \\ g_t = tanh(W_c * [\hat{H}_{t-1}, X_t] + b_c) \\ C_t = f_t \odot C_{t-1} + i_t \odot g_t \\ \hat{H}_t = o_t \odot tanh(C_t) \end{cases} \tag{2}$$

where SA represents the self-attention module, X_t and $\hat{H}_{t-1}$ represent the features that are aggregated through the self-attention module. Figure 2 shows the structure of the SA-ConvLSTM model, in which the self-attention memory module (SAM) is embedded in the ConvLSTM model. If the SAM module is removed, the SA-ConvLSTM model becomes a standard ConvLSTM model. The internal calculation formulas for the self-attention memory module are introduced in Section 3.2.2.

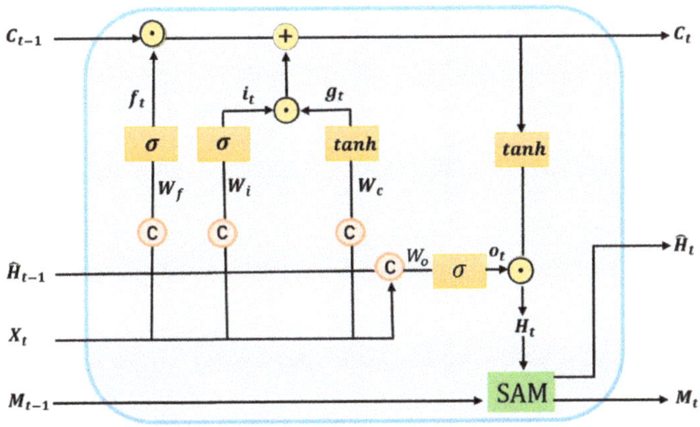

Figure 2. Structure of the SA-ConvLSTM model, where SAM is a self-attention memory module.

3.2.1. ConvLSTM Model

The ConvLSTM model consists of a memory unit and three memory gates. The memory unit can be considered to be a memory update gate. The structure of the memory unit is shown in Figure 3, and the equation is shown below:

$$\begin{cases} \bar{i}_t = \sigma\left(\overline{W}_i * \left[\hat{h}_{t-1}, \overline{X}_t\right] + \bar{b}_i\right) \\ \bar{f}_t = \sigma\left(\overline{W}_f * \left[\hat{h}_{t-1}, \overline{X}_t\right] + \bar{b}_f\right) \\ \bar{g}_t = tanh\left(\overline{W}_c * \left[\hat{h}_{t-1}, \overline{X}_t\right] + \bar{b}_c\right) \\ C_t = \bar{f}_t \odot C_{t-1} + \bar{i}_t \odot \bar{g}_t \\ \bar{o}_t = \sigma\left(\overline{W}_o * \left[\hat{h}_{t-1}, \overline{X}_t\right] + \bar{b}_o\right) \\ \hat{h}_t = \bar{o}_t \odot tanh(C_t) \end{cases} \tag{3}$$

where $\hat{h}_{t-1}$ denotes the hidden state at the previous moment and $\overline{X}_t$ denotes the current input value. σ and tanh are activation functions, * denotes the convolution operation, and ⊙ denotes the Hadamard product. $\bar{f}_t, \bar{i}_t, \bar{g}_t, \bar{o}_t$ denote forget gate, input gate, update gate, and output gate. $\overline{W}_f, \overline{W}_i, \overline{W}_c, \overline{W}_o$ and $\bar{b}_f, \bar{b}_i, \bar{b}_c, \bar{b}_o$ represent the weight matrix and bias matrix, respectively. C_{t-1} and C_t represents the unit state at the previous time and the unit state at the current time.

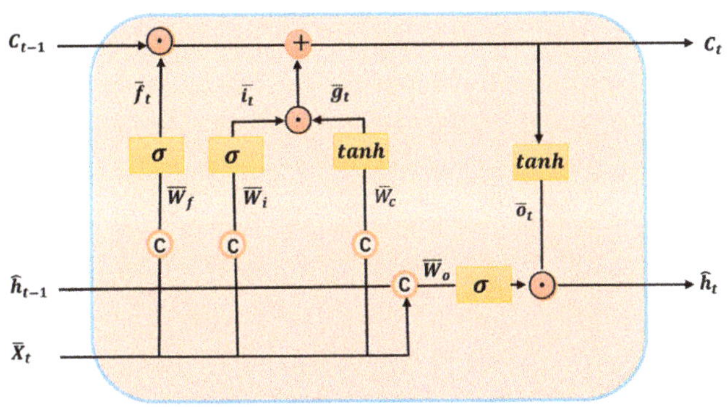

Figure 3. ConvLSTM memory unit structure.

3.2.2. Self-Attention Memory Module

The basic structure of the self-attention memory module (SAM) is presented in Figure 4, from which the SAM accepts two main inputs, namely the feature H_t of the current time step and the memory M_{t-1} of the previous step. It consists of three main parts: feature aggregation, memory updating, and output.

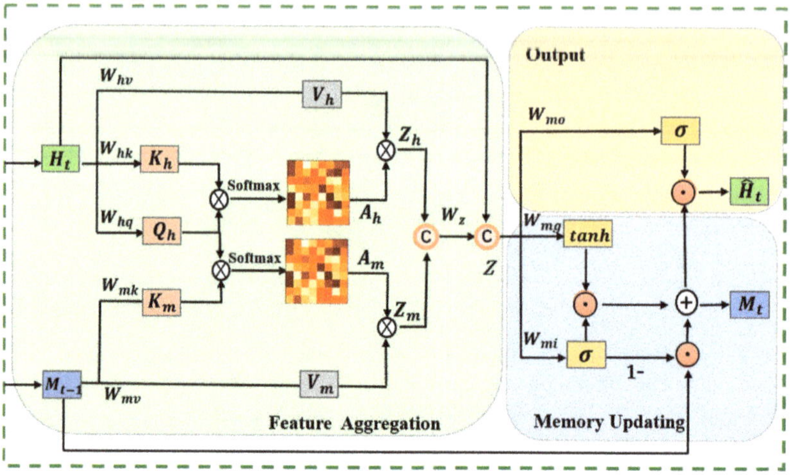

Figure 4. Structure of the Self-Attention Memory Module.

(1) Feature aggregation. The aggregated feature Z at each time step is formed by fusing Z_h and Z_m. Z_h is calculated by mapping the original feature H_t to different feature spaces as the query: $Q_h = W_{hq}H_t \in R^{\hat{C}\times N}$; the key: $K_h = W_{hk}H_t \in R^{\hat{C}\times N}$; and the value: $V_h = W_{hv}H_t \in R^{C\times N}$; where W_{hq}, W_{hk} and W_{hv} are the weight matrices of the 1×1 convolution kernel, C and $\hat{C}$ are the number of channels, and $N = H \times W$. Then, the similarity weight of the aggregated feature is calculated by the formula $A_h = Softmax\left(Q_h^T K_h\right)$ and finally the similarity scores of the aggregated feature are obtained by the formula $Z_h = V_h A_h$.The procedure for calculating Z_m is similar to that of Z_h, but the similarity score of the aggregated feature of Z_m is obtained by calculating the features of the current time step and the memory feature from the previous step. The final aggregated feature Z can be obtained with $Z = W_Z[Z_h; Z_m]$.

(2) Memory updating. The gating mechanism is used to update the memory M adaptively and enable SAM to capture long-range dependencies in the spatial and temporal domains. The aggregated feature Z and the original input H_t are used to generate the input gate i'_t and the fused feature g'_t. In addition, the forgetting gate is replaced with $1 - i'_t$ to reduce the parameters. The update process is shown below.

$$\begin{cases} i'_t = \sigma(W_{m;zi} * Z + W_{m;hi} * H_t + b_{m;i}) \\ g'_t = tanh\left(W_{m;zg} * Z + W_{m;hg} * H_t + b_{m;g}\right) \\ M_t = (1 - i'_t) \odot M_{t-1} + i'_t \odot g'_t \end{cases} \quad (4)$$

where $W_{m;zi}$, $W_{m;hi}$, $W_{m;hg}$ denote the weight matrices. $b_{m;i}$ and $b_{m;g}$ denote the bias matrices.

Compared with the original memory cell C in the ConvLSTM, which is updated only by convolutional operations, the proposed memory M is updated not only by convolutional operations but also by aggregation of features Z_t, obtaining global spatial correlations in time. Therefore, we consider that M_{t-1} can contain global past spatial-temporal information.

(3) Output. The output feature $\hat{H}_t$ of the self-attention memory module is a dot product between the output gate o'_t and the update memory M_t, with the following equation.

$$\begin{cases} o'_t = \sigma(W_{m;zo} * Z + W_{m;ho} * H_t + b_{m;o}) \\ \hat{H}_t = o'_t \odot M_t \end{cases} \tag{5}$$

where $W_{m;zo}$, $W_{m;ho}$ denote the weight matrix. $b_{m;o}$ denotes the bias matrix.

3.3. Model Construction

In this paper, the computational resources employed are Python 3.8. The Random Forest model was implemented via the Sklearn library, whereas the SA-ConvLSTM model was developed using the Tensorflow and Keras frameworks. Finally, the SST prediction model was constructed using both Keras and the Sklearn toolkit (Figure 5). The optimizing parameters of the Random Forest model have been set up in detail as follows: the number of decision trees is 1000, the maximum depth of the tree is 30, and the maximum number of features is 7. The SA-ConvLSTM model has three SA-ConvLSTM layers and one Conv3D layer with convolution kernels of (3, 3) and (3, 3, 3), respectively, and the optimization algorithm is Adam. In addition, to prevent overfitting, we have added dropout layers and set them to 0.4 and 0.3, respectively, and the maximum number of iterations set to 40.

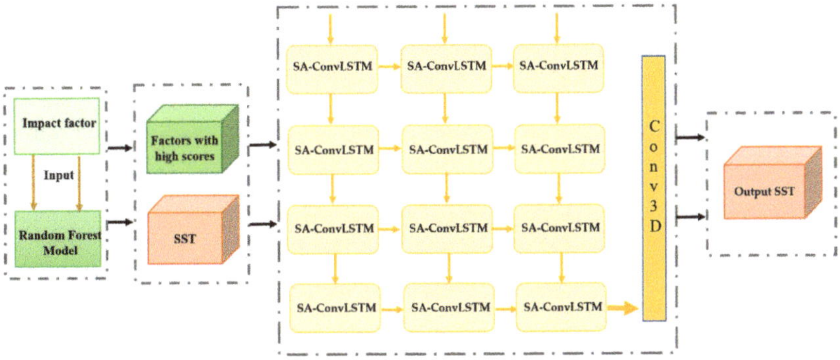

Figure 5. Structure of SST prediction based on Random Forest and SA-ConvLSTM model.

3.4. Evaluation Functions

To evaluate the SST prediction performance, this paper uses the root mean square error (*RMSE*), the mean absolute error (*MAE*), the absolute error (*AE*), the coefficient of determination (R^2), and the correlation coefficient (*R*) to compare and analyze the predicted SST with the actual SST data. The equations are as follows.

$$RMSE = \sqrt{\frac{1}{m}\sum_{i=1}^{m}\left(sst_o - sst_p\right)^2} \tag{6}$$

$$MAE = \frac{1}{m}\sum_{i=1}^{m}\left|sst_o - sst_p\right| \tag{7}$$

$$AE = \left|sst_o - sst_p\right| \tag{8}$$

$$R^2 = 1 - \frac{\sum_{i=1}^{m}\left(sst_o - sst_p\right)^2}{\sum_{i=1}^{m}\left(sst_o - \overline{sst_o}\right)^2} \tag{9}$$

$$R = \frac{\sum_{i=1}^{m}\left(var_i - \overline{var}\right)\left(sst_o - \overline{sst_o}\right)}{\sqrt{\sum_{i=1}^{m}\left(var_i - \overline{var}\right)^2}\sqrt{\sum_{i=1}^{m}\left(sst_o - \overline{sst_o}\right)^2}} \tag{10}$$

$$PBIAS = \frac{\sum_{i=1}^{m}(sst_p - sst_o)}{\sum_{i=1}^{m} sst_o} \times 100\% \quad (11)$$

where m is the total number of samples, sst_o and sst_p are the OSTIA data and the predicted value of the SST. The lower the value of RMSE, MAE, and AE, the more accurate the predicted result. The closer the R^2 is to 1, the better the fit between the predicted SST and the OSTIA data. The higher the R-value, the better the correlation. The Percent bias (PBIAS) measures the average tendency of the simulated data to be larger or smaller than observed counterparts. PBIAS can also help identify average model simulation bias (overprediction vs. underprediction) and incorporate measurement uncertainty.

To evaluate the improvement of the model using multi-element input dataset compared with a single-element input dataset, this paper uses the following formula:

$$Skill_{index} = \frac{\left|E_{index}^{ds} - E_{index}^{dm}\right|}{E_{index}^{ds}} \times 100\% \quad (12)$$

where E_{index}^{ds} represents the error index of the model's prediction using a single-element input dataset, E_{index}^{dm} represents the error index of the model's prediction using a multi-element input dataset, and the error index represents the RMSE value or MAE value.

4. Analysis of Results

4.1. Feature Importance Assessment for Random Forests

The SST in the ECS varies greatly from coastal to open sea, influenced by complex shelf circulations and monsoons. The mechanisms of SST variability are quite different in different regions, such as estuaries, coastal upwelling, and the Kuroshio Current. The mechanisms of SST variability are quite different in different regions, such as estuaries, coastal upwelling, and the Kuroshio Current. However, they are all influenced by atmospheric forcings such as the 2 m air temperature (T2m), sea surface solar shortwave radiation flux (*SWR*), sea surface longwave radiation flux (*LWR*), sea surface latent heat flux (LHF), sea surface sensible heat flux (*SHF*), U-component of 10 m wind and V-component of 10 m wind (10 m-wind). Since the data-driven methods could predict the future trend of SST based on a lot of SST data and ignore the physical mechanisms, not all these atmospheric forcings contribute equally to improving the accuracy of SST prediction. To select which kind of data could most improve the accuracy of SST, this study uses the Random Forest model to test the contribution of these data according to their feature importance assessment results in six typical locations L1~L6 (Figure 6). These sites are selected as follows: L1 at the Yangtze River estuary, where the SST has obvious seasonal variations. L2 and L3 are selected in the Kuroshio region, where temperature and salinity are influenced by the Kuroshio. L4 is selected in the northwest Pacific Ocean, which is far from the land, and the SST changes are relatively stable. L5 was chosen to be close to the Taiwan Strait, where temperature and salinity are influenced by the Taiwan warm current and the Min-Zhe coastal current [51,52]. L6 was chosen to be close to the Tsushima Strait, where temperature and salinity are influenced by the Tsushima Warm Current, a branch of the Kuroshio in the East China Sea region.

Figure 7 shows the prediction results of the Random Forest in 2020. In comparison with the OSTIA data, the Random Forest can accurately reproduce the long-term trend of the SST. The prediction results of Random Forest are not very accurate for the extremum, while all R^2 at six locations are above 94%, showing that the model built by Random Forest is reliable and stable. Both the R and PBIAS also show that the predicted SST result from Random Forest is well in agreement with the SST of OSTIA. The Random Forest feature importance score can then be used to select the key atmospheric forcings that affect SST variability. To enhance the accuracy of variable selection for predicting SST results, we compared both the Random Forest feature importance score and the correlation coefficient R (Figure 8). The results of the Random Forest forecast are more accurate when 2 m air temperature and sea surface longwave radiation are taken into account, with 2 m air temperature being the most significant factor (Figure 8a). Conversely, 10 m wind speed, sea surface solar shortwave radiation, and heat flux have a lesser impact on the accuracy of the results. The sea surface longwave radiation and the 2 m air temperature have a strong positive correlation with the SST, with R values around 0.8 (Figure 8b). On the other hand, the R-value of the 10 m wind speed is mostly around −0.2, indicating a weak negative correlation with the SST. Although the histograms of the characteristic importance scores of the Random Forest and the histograms of R values differ, it can be observed that they have higher scores for longwave radiation and air 2 m temperature and lower scores for 10 m wind speed. In Section 4.2 of this paper, we selected longwave radiation (LWR) and 2 m air temperature (T2m) as influencing factors on SA-ConvLSTM forecast results and constructed four different datasets based on OSTIA SST, LWR, and 2 m air temperature data.

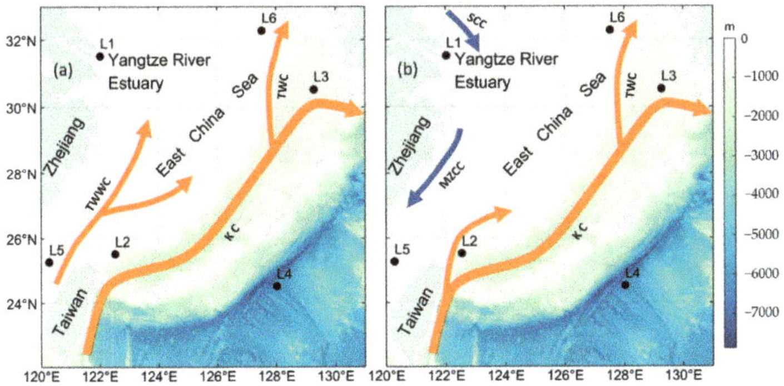

Figure 6. Schematic diagram of the East China Sea study area, where the black points are the six selected representative locations, where (**a**) shows the distribution of currents in summer and (**b**) the distribution of currents in winter [51,52]. SCC, MZCC, TWWC, TWC, and KC represent the Subei Coastal Current, the Min-Zhe (Fujian–Zhejiang) Coastal Current, the Taiwan Warm Current, the Tsushima Warm Current, and the Kuroshio. The blue line is the cold current, and the orange line is the warm current.

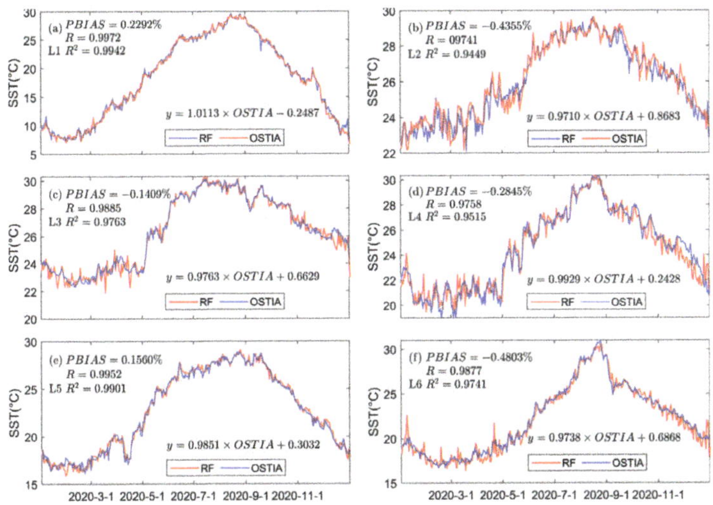

Figure 7. OSTIA data and prediction results of Random Forest (RF) for different locations, where (**a–f**) represent the sites of L1~L6, and the red line is the OSTIA data, the blue line is the prediction results of RF, and the horizontal coordinates represent the data.

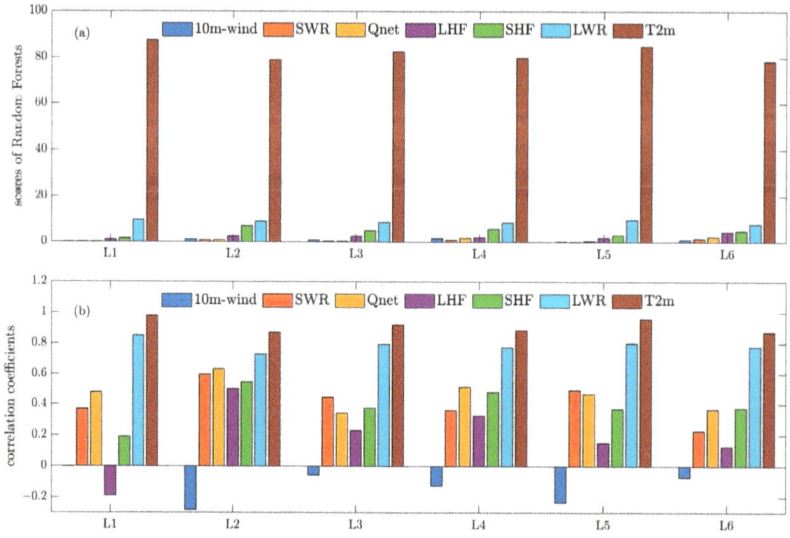

Figure 8. Characteristic importance assessment scores for Random Forests (**a**) and correlation coefficients (**b**) at six sites L1~L6. 10 m-wind is the wind speed of 10 m, SWR is the sea surface solar shortwave radiation, *Qnet* is the heat flux, LHF is the sea surface latent heat flux, SHF is the sea surface sensible heat flux, LWR is the sea surface longwave radiation and T2m is the 2 m air temperature.

4.2. Spatial Variation Analysis of Prediction Results from SA-ConvLSTM Models

Based on the results of the Random Forest selection and the correlation coefficient, four datasets were constructed as input to SA-ConvLSTM: SST-only, SST-T2m, SST-LWR, and SST-T2m-LWR. SST-only includes SST data only, SST-T2m includes both SST and 2 m air temperature data, SST-LWR includes SST and longwave radiation data, and SST-T2m-LWR includes SST, 2 m air temperature and longwave radiation data. To analyze the prediction accuracy of different input datasets, Figures 9–11 show the spatial variation of the seasonal mean predicted SST, AE, and RMSE. The four seasons are defined as follows: spring (March to May), summer (June to August), autumn (September to November), and winter (December to February). The predicted sea surface temperature (SST) of the four datasets has a spatial distribution similar to that of the OSITA data in all four seasons (Figure 9). The spatial characteristics indicate a general trend of low values in the northwest coastal seas and high values in the southeast open sea. The SA-ConvLSTM can represent the seasonal variation of SST in the ECS. The spatial distribution in spring and winter is consistent with OSTIA, while there are some differences in summer and autumn. Compared with OSTIA, the predicted SST by all four datasets in the Pacific Northwest region in summer and autumn is underestimated. The bias of the predicted result using the SST-only dataset is more apparent than the results of the other three datasets. The accuracy of the SST prediction can be improved by including T2m and LWR as additional input data.

The absolute error (AE) between predicted and observed values is low in spring and winter when using four different datasets as input, with most values around 0.2 °C (Figure 10). However, the AE between the predicted SST and OSTIA SST is largest in summer, regardless of which of the four datasets is used as input data, with an AE of about 1 °C in the Pacific Northwest region. The AE of all predicted SST in autumn is also evident in the Kuroshio region and coastal seas. The spatial distribution of AE differs significantly between the four seasons. During spring and winter, the AEs in coastal regions are higher than those in the open sea, indicating that the predicted temperature may be affected by coastal currents. During summer and autumn, the AE in the Kuroshio region is significantly higher than that in the coastal region, suggesting that the predicted temperature may be influenced by the Kuroshio.

Figure 11 shows the spatial distribution of the root mean square error (RMSE) of the predicted SST using the four datasets. During summer, the RMSE is highest throughout the study area, with the largest RMSE observed in the region adjacent to Taiwan Island. The RMSE is lowest in winter, but the Yangtze River estuary, Zhejiang coastal seas, and the northeast sea of Taiwan have much higher RMSE than other parts of the study area. The spatial distributions of the RMSE in autumn are similar to those in winter, while the RMSE is clearly demarcated by the Kuroshio in spring, with high values in the northwestern area of the Kuroshio and low values in the southeastern area of the Kuroshio. In summer, the RMSE of datasets SST-T2m and SST-T2m-LWR are lower than those of SST-only and SST-LWR, which indicates that datasets including 2 m air temperature data perform better than those without. All four datasets have larger RMSE values and lower prediction accuracy in the summer in the Kuroshio regions, indicating that there is a significant correlation between the spatial distribution of model prediction accuracy and ocean circulations.

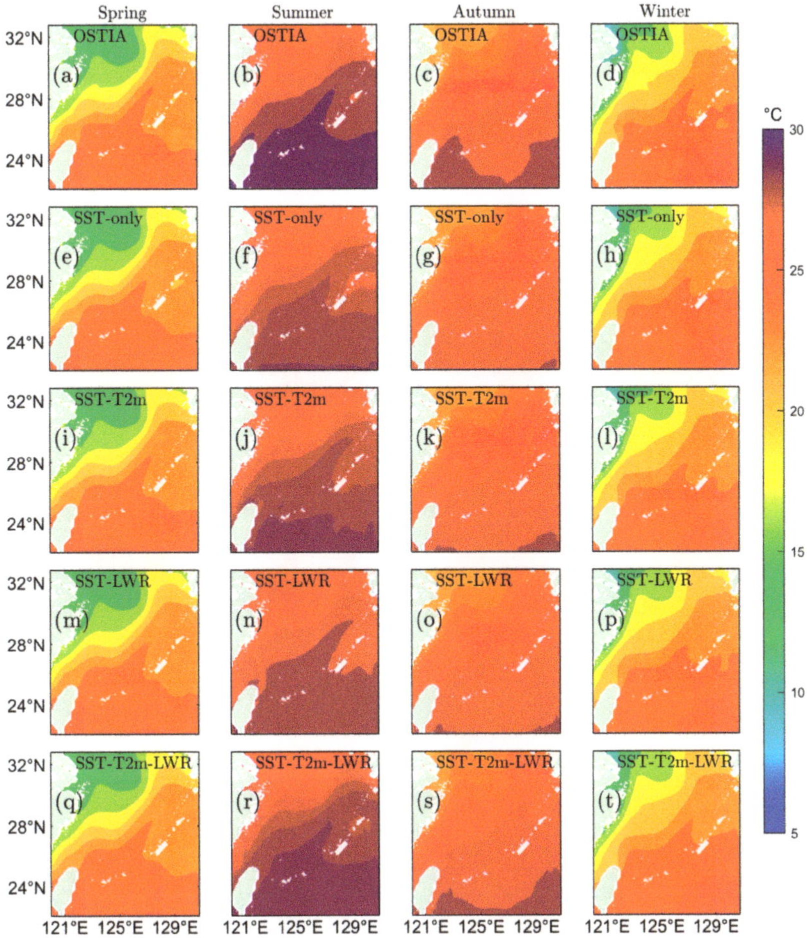

Figure 9. (a–t) Seasonal spatial distribution of OSTIA and the predicted SST from four constructed datasets: SST-only, SST-T2m, SST-LWR and SST-T2m-LWR.

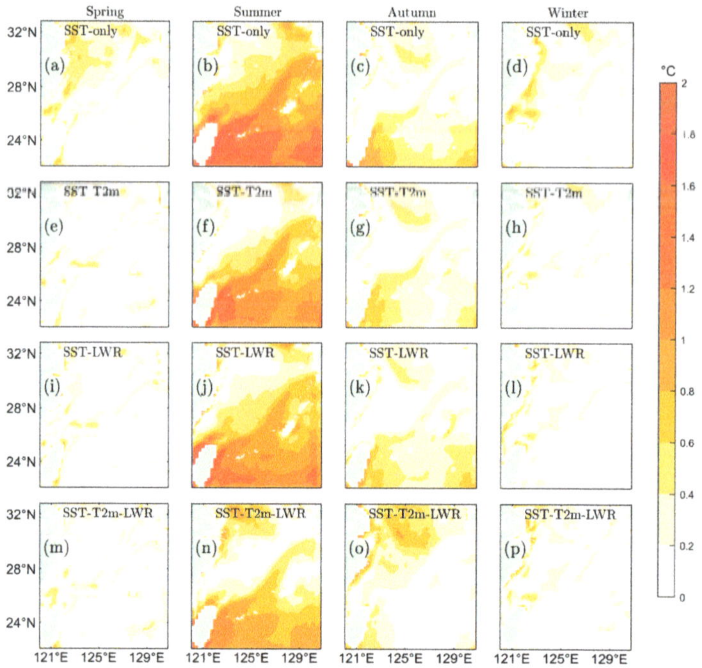

Figure 10. (**a–p**) Seasonal spatial distribution of AE of the predicted SST from four constructed datasets: SST-only, SST-T2m, SST-LWR and SST-T2m-LWR.

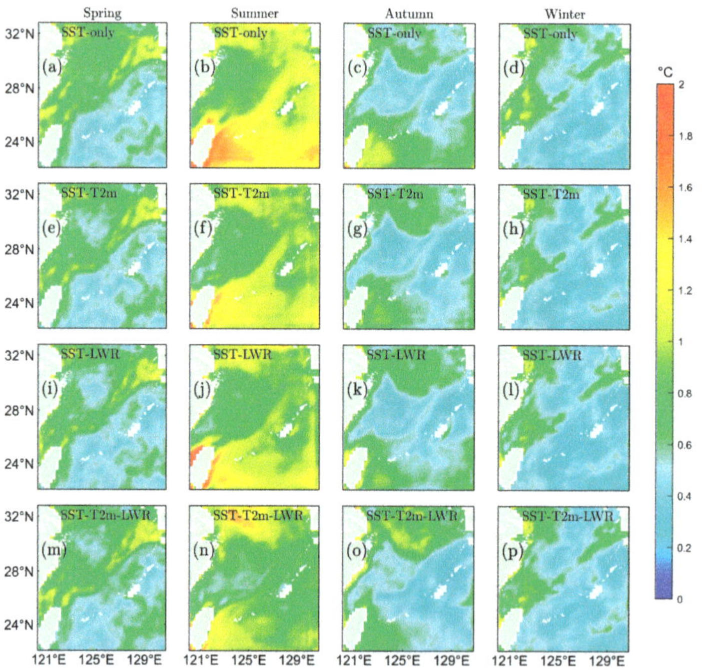

Figure 11. (**a–p**) Seasonal spatial distribution of RMSE of the predicted SST from four constructed datasets: SST-only, SST-T2m, SST-LWR and SST-T2m-LWR.

4.3. Temporal Variation Analysis of Prediction Results from SA-ConvLSTM Models

Table 1 presents the mean RMSE and MAE of the prediction results for four input datasets in 2020, along with the percentage improvement compared to the SST-only dataset. The RMSE and MAE of the prediction results using the SST-only dataset are the highest, with values of about 0.72 °C and 0.57 °C, respectively. Both the RMSE and MAE were reduced when using the SST-T2m, SST-LWR, and SST-T2m-LWR datasets. Compared to using only SST data as model training and input data for SA-ConvLSTM, adding T2m and LWR data separately can improve model performance. By adding the T2m and LWR data individually, the prediction skill of the model can be improved by about 9.9% and 9.43% for the RMSE and by about 8.97% and 8.21% for the MAE. However, when T2m and LWR data are added together, the prediction skill of the model using SST-T2m-LWR can be improved by 10.75% for RMSE and 9.06% for MAE. These results are slightly better than the results obtained by adding the T2m and LWR data separately. In descending order, both the RMSE and MAE of four datasets are SST-only, SST-LWR, SST-T2m, and SST-T2m-LWR. This shows that model prediction performance improves as the number of input datasets increases. Nevertheless, our experiments also show that when the number of input datasets increases from two to three types, there is no significant improvement in model performance. The Random Forest importance scores and R-value can be used as a reference for selecting additional input data to improve model prediction performance when using a data-driven method.

Table 1. The mean RMSE and MAE of four datasets in 2020 (unit: °C) and the percentage improvement.

	SST-Only	SST-T2m	SST-LWR	SST-T2m-LWR
MAE	0.5563	0.5064	0.5106	0.5059
RMSE	0.7221	0.6506	0.6540	0.6445
$Skill_{RMSE}$	0%	9.9%	9.43%	10.75%
$Skill_{MAE}$	0%	8.97%	8.21%	9.06%

To quantitatively assess the forecasting effectiveness of the model over the four seasons. Figure 12 displays the variation in the histogram of mean RMSE and MAE of prediction results for four input datasets across four seasons in 2020. In terms of seasonal variation, both the RMSE and MAE for the four datasets show similar seasonal variations, with the highest values in summer and the lowest values in winter. In summer, using the SST-only dataset resulted in the highest RMSE and MAE of about 1.01 °C and 0.88 °C, respectively, but using the SST-T2m-LWR dataset resulted in the lowest RMSE and MAE of about 0.82 °C and 0.70 °C, respectively. However, in winter, using the SST-only dataset resulted in the lowest RMSE and MAE of about 0.5 °C and 0.39 °C, respectively, but using the SST-T2m dataset resulted in the lowest RMSE and MAE of about 0.45 °C and 0.35 °C, respectively. In spring and autumn, the RMSE and MAE are about 0.6 °C and 0.5 °C, respectively. In terms of input datasets, the model's performance improvement due to T2m and LWR is most noticeable during the summer, with less significant improvements in other seasons. For the three seasons of spring, autumn, and winter, the RMSE differences among the four datasets are minimal. Similarly, the MAE differences among the four datasets are also small, except for summer.

Figure 13 shows the results of an analysis of the variation in the model's prediction errors using different input sets from month to month. The figure clearly shows that the maximum value of MAE and the minimum value of R^2 are obtained simultaneously. This relationship is particularly evident in summer. Furthermore, it was observed that the error curve in the prediction results increases as the month progresses, peaking in August when different input datasets are used. The model demonstrates a decrease in error as the month progresses after August. However, there is less improvement in prediction performance using the multi-input dataset compared to the single-input dataset before June and after October. In individual months, the model's predictions using two datasets are slightly better than the model's predictions using three input datasets. For example, in December, the model's predictions using the SST-only input dataset have lower MAE values than the model's predictions using the SST-T2m-LWR input dataset. From June to October, the MAE value gradually decreases as the number of input datasets increases. This indicates that the model's prediction performance improves with more input datasets.

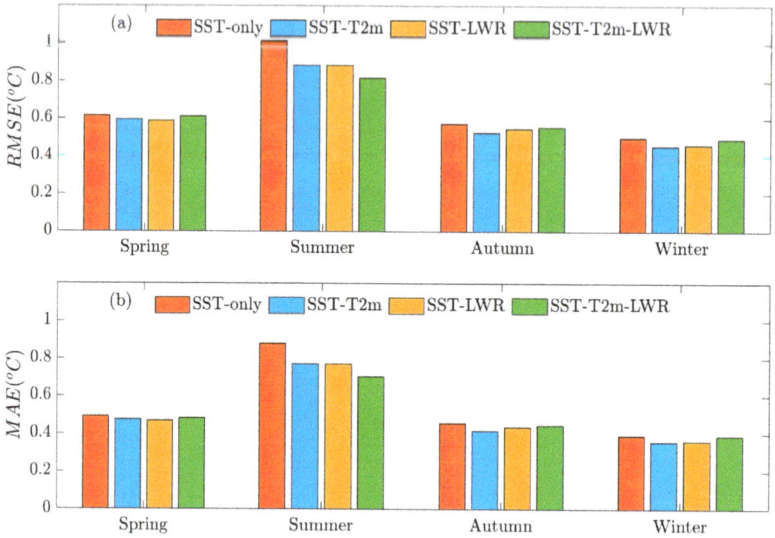

Figure 12. Seasonal variation of RMSE (**a**) and MAE (**b**) of prediction SST of four constructed datasets: SST-only, SST-T2m, SST-LWR and SST-T2m-LWR.

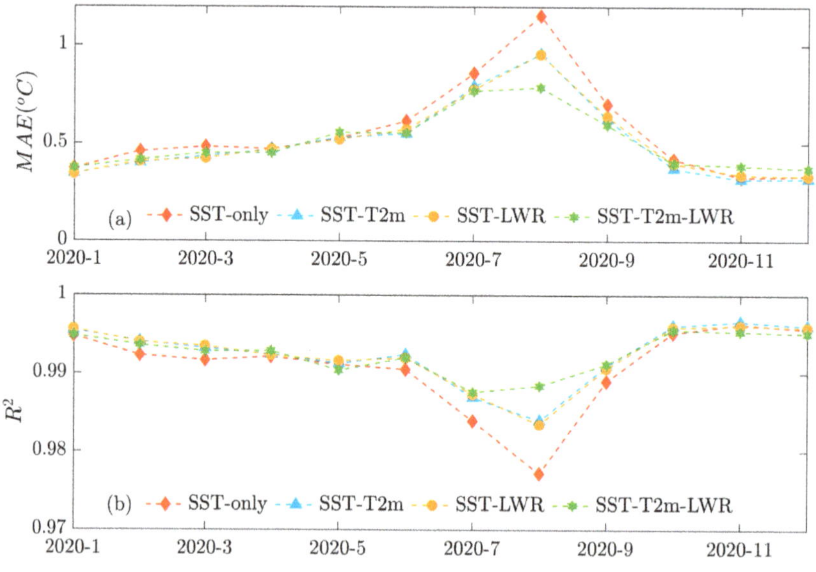

Figure 13. Seasonal monthly variation of MAE (**a**) and R^2 (**b**) of prediction SST of four constructed datasets: SST-only, SST-T2m, SST-LWR and SST-T2m-LWR.

5. Discussion

Numerical ocean modeling and data-driven methods are the two main methods for predicting sea temperature. It is important to note that both methods have their advantages and disadvantages. Numerical ocean modeling considers comprehensive physical mechanisms and employs various methods to reduce input errors and improve forecast accuracy. On the other hand, data-driven methods rely on observed data to improve forecast accuracy without considering physical mechanisms.

Considering extensive physical mechanisms and using numerical algorithms to solve governing equations, ocean modeling is widely used to simulate or predict the ocean state [53]. These traditional numerical ocean models are usually computationally expensive and require various input data as complex geometric boundaries. Thus, accurate prediction of SST using a numerical ocean model remains a challenge due to the complex dynamical and thermal processes at the air–sea interface, such as ocean waves [54], turbulence, and radiation fluxes [55]. In general, accurate SST prediction by numerical ocean models usually requires input data, including topography, initial temperature and salinity fields, bottom and lateral boundary conditions, and atmospheric forcing fields, and it usually improves accuracy by data assimilation. For example, Lin used POM to predict the three-dimensional temperature field in the Bohai Sea, Yellow Sea, and East China Sea [56]. The Levitus climate monthly mean temperature and salinity were used as the initial fields, and model results from MOM2 were used as boundary conditions, respectively. The nudging assimilation method was also used to improve the accuracy of temperature prediction. As a result, the RMS between the temperature simulated by heat flux and the observed data are 1.56 °C, 0.88 °C, and 1.23 °C, respectively [57]. The atmospheric data and model parameter scheme also affected the simulation result even though the model have the same initial and boundary conditions. Gao et al. used HYCOM to simulate the sea surface temperature of the Tropical and North Pacific [13]. Based on the different combinations of two air–sea flux data sets (COADS and ECMWF) and two bulk parameter formulas (non-constant and constant), they found that the largest difference is about 4 °C using different combinations [13]. Compared with numerical ocean models, the topography, the boundary condition, and the physical parameter schemes can be ignored or disregarded by data-driven methods.

Many Different data-driven methods are used to predict the SST in the East China Sea [16,17,34,57–59]. However, this data-driven method only uses a single SST as input data and does not consider the influence of other factors affecting the sea surface temperature on the model prediction. The model proposed in this paper has preliminarily explored the influence of atmospheric factors on SST prediction. The SA-ConvLSTM used in this paper achieved good prediction results using the SST-only dataset, with the RMSE and MAE of about 0.72 °C and 0.56 °C, respectively. The atmospheric data for T2m and LWR are not used as model forcing conditions but as training and input data. Both the SST-T2m and SST-LWR datasets can improve model performance, but when the combined SST-T2m-LWR dataset is used, the prediction skill of the model improves significantly, with a 10.75% reduction in RMSE and a 9.06% reduction in MAE. The SA-ConvLSTM has the largest error in summer and the lowest error in other seasons. It indicated that the addition of T2m and LWR significantly improves the model's performance in summer while only slightly improving it in other seasons. The experiment results indicate that the accuracy of predicting the SST can be improved using more related data in the data-driven method.

This study shows that the Random Forest importance scores and R-value would help to select key influence factors to improve model performance. However, further exploration is needed to understand why the model has a larger MAE and RMSE in summer compared to the other three seasons. There are several possible solutions to this problem. Expanding the temporal span of data would be one solution way. The data used in this paper only covers the period from 2010 to 2020 and does not include signals of long-term trends, such as PDO cycles or global warming. It is necessary to experimentally test whether the model can be more accurate when using much longer data, such as 30 years or more.

Adjusting the parameters and settings of the model is another way to improve prediction accuracy. For example, modifying the length of input data can lead to more reasonable results. Additionally, integrating data-driven methods with the physical mechanisms that affect SST could be a potential solution, which requires further exploration.

6. Conclusions

This study aims to predict SST in the East China Sea using the Random Forest and SA-ConvLSTM model. The contribution of several key atmospheric forcings to the SST prediction has been tested. Based on the Random Forest importance assessment and the correlation coefficient R, the 2 m air temperature and longwave radiation were identified as the two most important additional data that can affect the prediction results of the SA-ConvLSTM model. Four datasets were constructed as input to SA-ConvLSTM: SST-only, SST-T2m, SST-LWR, and SST-T2m-LWR. Using all four datasets can represent the SST in the ECS well while having the highest RMSE and AE in summer and the lowest RMSE and AE in winter.

Compared with the prediction results using the SST-only dataset, when using the SST-T2m and SST-LWR, the prediction skill of the model can be improved by about 9.9% and 9.43% for the RMSE and by about 8.97% and 8.21% for the MAE, respectively. Using the SST-T2m-LWR dataset, the model's prediction skill can be improved by 10.75% for RMSE and 9.06% for MAE. Based on the Random Forest and the correlation coefficient R, the selection of the two main atmospheric forces can significantly improve the model's prediction performance. The model shows higher MAEs and RMSEs in summer compared to the other three seasons. More data with a longer time span, adjustment of model parameters and settings, and integration of data-driven methods with physical mechanisms could be used in further studies.

Author Contributions: Conceptualization, X.J., Y.W., X.L. and Q.J.; methodology, X.J. and Q.J.; software, X.J.; validation, X.J., Q.J. and L.J.; formal analysis, X.J., Q.J. and M.X.; investigation, X.J. and Q.J.; resources, Q.J. and X.L.; data curation, M.X., Z.M., X.L. and X.J; writing—original draft preparation, X.J.; writing—review and editing, Q.J. and X.J.; visualization, X.J.; supervision, Q.J.; project administration, Q.J. and L.J.; funding acquisition, Q.J. and L.J. All authors have read and agreed to the published version of the manuscript.

Funding: This research was funded by the construction Project of China ASEAN (Association of Southeast Asian Nations) blue partnership (No.99950410), National Natural Science Foundation of China under grant No. 41806004, and the Innovation Group Project of Southern Marine Science and Engineering Guangdong Laboratory (Zhuhai) (SML2020SP007 and 311020004).

Institutional Review Board Statement: Not applicable.

Informed Consent Statement: Not applicable.

Data Availability Statement: OSTIA data sets can be found here: https://www.ncei.noaa.gov/data/oceans/ghrsst/L4/GLOB/UKMO/OSTIA/ (accessed on 11 November 2021). ERA5 data sets can be found here: https://cds.climate.copernicus.eu/ (accessed on 20 December 2021).

Conflicts of Interest: The authors declare no conflicts of interest.

References

1. Herbert, T.D.; Peterson, L.C.; Lawrence, K.T.; Liu, Z. Tropical Ocean Temperatures over the Past 3.5 Million Years. *Science* **2010**, *328*, 1530–1534. [CrossRef] [PubMed]
2. Bouali, M.; Sato, O.T.; Polito, P.S. Temporal trends in sea surface temperature gradients in the South Atlantic Ocean. *Remote Sens. Environ.* **2017**, *194*, 100–114. [CrossRef]
3. Yao, S.-L.; Luo, J.-J.; Huang, G.; Wang, P. Distinct global warming rates tied to multiple ocean surface temperature changes. *Nat. Clim. Chang.* **2017**, *7*, 486–491. [CrossRef]
4. Lin, X.; Shi, J.; Jiang, G.; Liu, Z. The influence of ocean waves on sea surface current field and sea surface temperature under the typhoon background. *Mar. Sci. Bull.* **2018**, *37*, 396–403. [CrossRef]
5. Nagasoe, S.; Kim, D.-I.; Shimasaki, Y.; Oshima, Y.; Yamaguchi, M.; Honjo, T. Effects of temperature, salinity and irradiance on the growth of the red tide dinoflagellate Gyrodinium instriatum Freudenthal et Lee. *Harmful Algae* **2006**, *5*, 20–25. [CrossRef]

6. Li, S.; Goddard, L.; DeWitt, D.G. Predictive Skill of AGCM Seasonal Climate Forecasts Subject to Different SST Prediction Methodologies. *J. Clim.* **2008**, *21*, 2169–2186. [CrossRef]
7. Solanki, H.U.; Bhatpuria, D.; Chauhan, P. Signature analysis of satellite derived SSHa, SST and chlorophyll concentration and their linkage with marine fishery resources. *J. Mar. Syst.* **2015**, *150*, 12–21. [CrossRef]
8. Jiao, N.; Zhang, Y.; Zeng, Y.; Gardner, W.D.; Mishonov, A.V.; Richardson, M.J.; Hong, N.; Pan, D.; Yan, X.-H.; Jo, Y.-H.; et al. Ecological anomalies in the East China Sea: Impacts of the Three Gorges Dam? *Water Res.* **2007**, *41*, 1287–1293. [CrossRef] [PubMed]
9. Patil, K.; Deo, M.C. Prediction of daily sea surface temperature using efficient neural networks. *Ocean Dyn.* **2017**, *67*, 357–368. [CrossRef]
10. Wei, X.; Xiang, Y.; Wu, H.; Zhou, S.; Sun, Y.; Ma, M.; Huang, X. AI-GOMS: Large AI-Driven Global Ocean Modeling System. *arXiv* **2023**, arXiv:2308.03152. [CrossRef]
11. Stockdale, T.N.; Balmaseda, M.A.; Vidard, A. Tropical Atlantic SST Prediction with Coupled Ocean–Atmosphere GCMs. *J. Clim.* **2006**, *19*, 6047–6061. [CrossRef]
12. Krishnamurti, T.; Chakraborty, A.; Krishnamurti, R.; Dewar, W.; Clayson, C. Seasonal Prediction of Sea Surface Temperature Anomalies Using a Suite of 13 Coupled Atmosphere–Ocean Models. *J. Clim.* **2006**, *19*, 6069–6088. [CrossRef]
13. Song, G.; Xianqing, L.; Wang, H. Sea Surface Temperature Simulation of Tropical and North Pacific Basins Using a Hybrid Coordinate Ocean Model (HYCOM). *Mar. Sci. Bull.* **2008**, *10*, 1–14.
14. Zhang, X.; Zhang, W.; Li, Y. Characteristics of the sea temperature in the North Yellow Sea. *Mar. Forecast.* **2015**, *32*, 89–97.
15. Qian, Z.; Tian, J.; Cao, C.; Wang, Q. The numerical simulation and the assimilation technique of the current and the temperature field in the Bohai Sea, the Huanghai Sea and the East China Sea. *Haiyang Xuebao* **2005**, *27*, 1–6.
16. Xiao, C.; Chen, N.; Hu, C.; Wang, K.; Xu, Z.; Cai, Y.; Xu, L.; Chen, Z.; Gong, J. A spatiotemporal deep learning model for sea surface temperature field prediction using time-series satellite data. *Environ. Model. Softw.* **2019**, *120*, 104502. [CrossRef]
17. Wie, L.; Guan, L.; Qu, L.; Li, L. Prediction of Sea Surface Temperature in the South China Sea by Artificial Neural Networks. In Proceedings of the IGARSS 2019—2019 IEEE International Geoscience and Remote Sensing Symposium, Yokohama, Japan, 28 July–2 August 2019; pp. 8158–8161. [CrossRef]
18. Garcia-Gorriz, E.; Garcia-Sanchez, J. Prediction of sea surface temperatures in the western Mediterranean Sea by neural networks using satellite observations. *Geophys. Res. Lett.* **2007**, *34*, L11603. [CrossRef]
19. Patil, K.; Deo, M.C. Basin-Scale Prediction of Sea Surface Temperature with Artificial Neural Networks. *J. Atmos. Ocean. Technol.* **2018**, *35*, 1441–1455. [CrossRef]
20. Zheng, G.; Li, X.; Zhang, R.H.; Liu, B. Purely satellite data-driven deep learning forecast of complicated tropical instability waves. *Sci. Adv.* **2020**, *6*, eaba1482. [CrossRef]
21. Xue, Y.; Leetmaa, A. Forecasts of tropical Pacific SST and sea level using a Markov model. *Geophys. Res. Lett.* **2000**, *27*, 2701–2704. [CrossRef]
22. Laepple, T.; Jewson, S.; Meagher, J.; O'Shay, A.; Penzer, J. Five year prediction of Sea Surface Temperature in the Tropical Atlantic: A comparison of simple statistical methods. *arXiv* **2007**, arXiv:physics/0701162. [CrossRef]
23. Kug, J.-S.; Kang, I.-S.; Lee, J.-Y.; Jhun, J.-G. A statistical approach to Indian Ocean sea surface temperature prediction using a dynamical ENSO prediction. *Geophys. Res. Lett.* **2004**, *31*, L09212. [CrossRef]
24. Peng, Y.; Wang, Q.; Yuan, C.; Lin, K. Review of Research on Data Mining in Application of Meteorological Forecasting. *J. Arid Meteorol.* **2015**, *33*, 19–27. Available online: http://www.ghqx.org.cn/EN/10.11755/j.issn.1006-7639(2015)-01-0019 (accessed on 13 October 2022).
25. Chaudhari, S.; Balasubramanian, R.; Gangopadhyay, A. Upwelling Detection in AVHRR Sea Surface Temperature (SST) Images using Neural-Network Framework. In Proceedings of the IGARSS 2008—2008 IEEE International Geoscience and Remote Sensing Symposium, Boston, MA, USA, 7–11 July 2008; pp. IV-926–IV-929. [CrossRef]
26. Wu, Z.; Jiang, C.; Conde, M.; Deng, B.; Chen, J. Hybrid improved empirical mode decomposition and BP neural network model for the prediction of sea surface temperature. *Ocean Sci.* **2019**, *15*, 349–360. [CrossRef]
27. Tangang, F.T.; Hsieh, W.W.; Tang, B. Forecasting regional sea surface temperatures in the tropical Pacific by neural network models, with wind stress and sea level pressure as predictors. *J. Geophys. Res. Oceans* **1998**, *103*, 7511–7522. [CrossRef]
28. Wu, A.; Hsieh, W.W.; Tang, B. Neural network forecasts of the tropical Pacific sea surface temperatures. *Neural Netw.* **2006**, *19*, 145–154. [CrossRef] [PubMed]
29. Tripathi, K.C.; Das, M.L.; Sahai, A. Predictability of sea surface temperature anomalies in the Indian Ocean using artificial neural networks. *Indian J. Mar. Sci.* **2006**, *35*, 210–220.
30. Aparna, S.G.; D'Souza, S.; Arjun, N.B. Prediction of daily sea surface temperature using artificial neural networks. *Int. J. Remote Sens.* **2018**, *39*, 4214–4231. [CrossRef]
31. Gupta, S.M.; Malmgren, B. Comparison of the accuracy of SST estimates by artificial neural networks (ANN) and other quantitative methods using radiolarian data from the Antarctic and Pacific Oceans. *Earth Sci. India* **2009**, *2*, 52–75.
32. Hou, S.; Li, W.; Liu, T.; Zhou, S.; Guan, J.; Qin, R.; Wang, Z. MIMO: A Unified Spatio-Temporal Model for Multi-Scale Sea Surface Temperature Prediction. *Remote Sens.* **2022**, *14*, 2371. [CrossRef]
33. Zhang, Q.; Wang, H.; Dong, J.; Zhong, G.; Sun, X. Prediction of Sea Surface Temperature Using Long Short-Term Memory. *IEEE Geosci. Remote Sens. Lett.* **2017**, *14*, 1745–1749. [CrossRef]

34. Xiao, C.; Chen, N.; Hu, C.; Wang, K.; Gong, J.; Chen, Z. Short and mid-term sea surface temperature prediction using time-series satellite data and LSTM-AdaBoost combination approach. *Remote Sens. Environ.* **2019**, *233*, 111358. [CrossRef]
35. He, Q.; Hu, Z.; Xu, H.; Song, W.; Du, Y. Sea Surface Temperature Prediction Method Based on Empirical Mode Decomposition-Gated Recurrent Unit Model. *Laser Optoelectron. Prog.* **2021**, *58*, 9. [CrossRef]
36. Ham, Y.-G.; Kim, J.-H.; Luo, J.-J. Deep learning for multi year ENSO forecasts. *Nature* **2019**, *573*, 568–572. [CrossRef] [PubMed]
37. Shi, X.; Chen, Z.; Wang, H.; Yeung, D.-Y.; Wong, W.-K.; Woo, W.-C. Convolutional LSTM Network: A machine learning approach for precipitation nowcasting. In Proceedings of the 28th International Conference on Neural Information Processing Systems, Montreal, QC, Canada, 7–12 December 2015; Volume 1, pp. 802–810. [CrossRef]
38. Zhang, K.; Geng, X.; Yan, X.H. Prediction of 3-D Ocean Temperature by Multilayer Convolutional LSTM. *IEEE Geosci. Remote Sens. Lett.* **2020**, *17*, 1303–1307. [CrossRef]
39. Luo, W.; Li, Y.; Urtasun, R.; Zemel, R. Understanding the effective receptive field in deep convolutional neural networks. In Proceedings of the 30th International Conference on Neural Information Processing Systems, Barcelona, Spain, 5–10 December 2016; pp. 4905–4913. [CrossRef]
40. Lin, Z.; Li, M.; Zheng, Z.; Cheng, Y.; Yuan, C. Self-Attention ConvLSTM for Spatiotemporal Prediction. *Proc. AAAI Conf. Artif. Intell.* **2020**, *34*, 11531–11538. [CrossRef]
41. Donlon, C.J.; Martin, M.; Stark, J.; Roberts-Jones, J.; Fiedler, E.; Wimmer, W. The Operational Sea Surface Temperature and Sea Ice Analysis (OSTIA) system. *Remote Sens. Environ.* **2012**, *116*, 140–158. [CrossRef]
42. Stewart, R.H. *Introduction to Physical Oceanography*; Prentice Hall: Upper Saddle River, NJ, USA, 2008; pp. 51–52. [CrossRef]
43. Breiman, L. Bagging Predictors. *Mach. Learn.* **1996**, *24*, 123–140. [CrossRef]
44. Breiman, L. Random Forests. *Mach. Learn.* **2001**, *45*, 5–32. [CrossRef]
45. Belgiu, M.; Drăguţ, L.D. Random forest in remote sensing: A review of applications and future directions. *ISPRS J. Photogramm. Remote Sens.* **2016**, *114*, 24–31. [CrossRef]
46. Chen, S.; Yu, L.; Zhang, C.; Wu, Y.; Li, T. Environmental impact assessment of multi-source solid waste based on a life cycle assessment, principal component analysis, and random forest algorithm. *J. Environ. Manag.* **2023**, *339*, 117942. [CrossRef] [PubMed]
47. Morse-McNabb, E.M.; Hasan, M.F.; Karunaratne, S. A Multi-Variable Sentinel-2 Random Forest Machine Learning Model Approach to Predicting Perennial Ryegrass Biomass in Commercial Dairy Farms in Southeast Australia. *Remote Sens.* **2023**, *15*, 2915. [CrossRef]
48. Wang, Y.; Chen, X.; Wang, Z.; Gao, M.; Wang, L. Integrating SBAS-InSAR and Random Forest for Identifying and Controlling Land Subsidence and Uplift in a Multi-Layered Porous System of North China Plain. *Remote Sens.* **2024**, *16*, 830. [CrossRef]
49. Su, H.; Li, W.; Yan, X.-H. Retrieving temperature anomaly in the global subsurface and deeper ocean from satellite observations. *J. Geophys. Res. Oceans* **2018**, *123*, 399–410. [CrossRef]
50. Cui, H.; Tang, D.; Mei, W.; Liu, H.; Sui, Y.; Gu, X. Predicting tropical cyclone-induced sea surface temperature responses using machine learning. *Geophys. Res. Lett.* **2023**, *50*, e2023GL104171. [CrossRef]
51. Chen, C.-T.A.; Sheu, D.D. Does the Taiwan Warm Current originate in the Taiwan Strait in wintertime? *J. Geophys. Res. Oceans* **2006**, *111*, C04005. [CrossRef]
52. Chen, C.-T.A. Rare northward flow in the Taiwan Strait in winter: A note. *Cont. Shelf Res.* **2003**, *23*, 387–391. [CrossRef]
53. Sonnewald, M.; Lguensat, R.; Jones, D.C.; Dueben, P.D.; Brajard, J.; Balaji, V. Bridging observations, theory and numerical simulation of the ocean using machine learning. *Environ. Res. Lett.* **2021**, *16*, 073008. [CrossRef]
54. Qiao, F.; Yuan, Y.; Yang, Y.; Zheng, Q.; Xia, C.; Ma, J. Wave-induced mixing in the upper ocean: Distribution and application to a global ocean circulation model. *Geophys. Res. Lett.* **2004**, *31*, L11303. [CrossRef]
55. Huang, C.J.; Qiao, F.; Chen, S.; Xue, Y.; Guo, J. Observation and Parameterization of Broadband Sea Surface Albedo. *J. Geophys. Res. Oceans* **2019**, *124*, 4480–4491. [CrossRef]
56. Lin, J. Numerical Simulation of Three-Dimensional Current Field and Temperature Field of the Bohai Sea, Yellow Sea and the East China Sea. Master's Thesis, Ocean University of China, Qingdao, China, 2004. Available online: https://www.dissertationtopic.net/doc/1278630 (accessed on 18 January 2024).
57. Jia, X.; Ji, Q.; Han, L.; Liu, Y.; Han, G.; Lin, X. Prediction of Sea Surface Temperature in the East China Sea Based on LSTM Neural Network. *Remote Sens.* **2022**, *14*, 3300. [CrossRef]
58. Xu, S.; Dai, D.; Cui, X.; Yin, X.; Jiang, S.; Pan, H.; Wang, G. A deep leaning approach to predict sea surface temperature based on multiple modes. *Ocean Modell.* **2023**, *181*, 102158. [CrossRef]
59. Yu, X.; Shi, S.; Xu, L.; Liu, Y.; Miao, Q.; Sun, M. A Novel Method for Sea Surface Temperature Prediction Based on Deep Learning. *Math. Probl. Eng.* **2020**, *2020*, 6387173. [CrossRef]

atmosphere

MDPI

Article

Response of Sea Surface Temperature and Chlorophyll-a to Typhoon Lekima (2019)

Yaowei Shi [1], Biyun Guo [1,2,*], Yuqian Niu [3,4], Venkata Subrahmanyam Mantravadi [5,*], Jushang Wang [1], Zhaokang Ji [1], Yingliang Che [1] and Menglu Ye [1]

1 Marine Science and Technology College, Zhejiang Ocean University, Zhoushan 316022, China
2 State Key Laboratory of Hydroscience and Engineering, Tsinghua University, Beijing 100084, China
3 Ocean College, Zhejiang University, Zhoushan 316022, China
4 State Key Laboratory of Satellite Ocean Environment Dynamics, Second Institute of Oceanography, Ministry of Natural Resources, Hangzhou 310012, China
5 School of Marine Science and Technology, Hainan Tropical Ocean University, Sanya 572022, China
* Correspondence: biyunguo@zjou.edu.cn (B.G.); mvsm.au@gmail.com (V.S.M.)

Abstract: Typhoon (hurricane) is the most influential process of ocean–air interaction on the synoptic scale; it has a great influence on the heat exchange, mixing and ecological processes in the upper ocean, which in turn affect sea surface temperature (SST), leading to chlorophyll-a (Chl-a) concentration variation. SST is also an important factor affecting marine fishery resources. Chl-a is closely related to the marine ecosystem and primary productivity. In this study, we analyzed the response of SST and Chl-a to Typhoon Lekima (2019) process. The result indicates that the response of temperature to typhoon decreases from the center to the outer edge, which has a good correlation with the location, path and influence area of the typhoon center. The mean SST in the study area (14°~40° N, 116°~136° E) decreased during the typhoon's passage, from 28.97 °C at the beginning (5 August) to 28.22 °C (15 August). The concentration of Chl-a was high in the northwest and coastal areas; its mean value in the study area decreased from 2 to 8 August (on 2 and 8 August, the concentration was 0.484 mg/m^3 and 0.405 mg/m^3, respectively). From 8 to 14 August, Chl-a decreased with the increase in SST, and 10 and 14 August were the two peak values of Chl-a (while SST was low). Chl-a concentration increased after the typhoon's landfall (from 15 to 31 August); the Chl-a trend was the same as that of SST. The stronger the typhoon and the longer the residence time, the greater the contribution to the increase in Chl-a concentration at sea surface. Typhoon-induced rainfall over the ocean surface, increased evaporation of seawater, enhanced mixing within the mixed layer and upwelling of the pycnocline resulted in an increase in Chl-a quantity. This study describes the spatial response of the upper ocean to typhoons. It provides a general method for the comprehensive assessment of typhoons in marginal seas and upper open oceans, which has wide applicability and good scientific application prospects.

Keywords: typhoon Lekima; SST; chlorophyll-a concentration; wind

1. Introduction

In the past few decades, typhoon frequency and intensity have increased, and its translation speed has slowed down [1], which has a significant impact on the regional marine environment. When a typhoon passes, the sea surface wind increases, and the mixing intensifies; therefore, the mixing layer usually deepens [2]. The cyclonic wind stress on the sea surface caused by the typhoon will lead to the upwelling of the lower seawater within a radius of several kilometers, resulting in the uplift of the thermocline [3]. The change in the thermocline will last for several inertial oscillation cycles after the typhoon [4]. After the typhoon, the heat flux from the ocean to the atmosphere is as high as 100 W/m^2, and there is a cooling process in the ocean mixing layer, which makes the temperature of the surface mixing layer drop by 1 °C on average [5,6]. The influence of typhoon on the ocean can exceed 1000 m depth; the temperature decreases with depth, the stratification is obvious at depth, and the exchange is weak at the bottom [7–11]. The stratification of temperature hindered the transport of nutrients from the lower layer to the mixing layer, resulting in lower nutrients in the surface layer and higher Chl-a concentration in the lower layer [12]. Typhoon caused strong mixing of seawater, bringing the lower layer cold water to the surface layer [13]. The spatial distribution of Chl-a depends on the relationship among the thickness of the mixed layer, the trophic thermocline and the maximum Chl-a concentration in the lower layer [12].

The responses of Chl-a and SST to typhoons are very different both in space and between single typhoon events [14]. Typhoon Lekima formed on 4 August 2019. In this study, the response of SST and Chl-a around the track of Lekima to the typhoon process was studied. Chl-a is important for the study of primary productivity in the marine ecosystem, which provides an important basis for the study of carbon cycle in the marine atmospheric system and global heat effect circulation, and it is of great significance for the study of ocean current (upwelling, coastal current, etc.) and marine fishery management [15–17]. Satellite remote-sensing data (such as MODIS, SeaWiFS, Goci, MERIS, etc.) have high spatial-temporal resolution [7] and can be used for large-area synchronous observation; therefore, they are widely used in marine Chl-a concentration estimation and spatial-temporal dynamic analysis [15–17]. SST combined with other remote-sensing data, such as chlorophyll a, wind field and sea surface height, has increasingly become an important data source in oceanographic research, especially in severe weather conditions, such as typhoons. Due to the severe weather of typhoon, it is difficult to measure the condition of the ocean by conventional methods.

The development of remote-sensing technology provides an effective means for the study of such extreme events. There are many studies on the use of satellite data to detect the ocean surface characteristics over the area of typhoon passage, while there are few studies on the Chl-a concentration, temperature and the relationship between them before and after the typhoon [18–21]. In this study, remote-sensing technology was used to study the temporal and spatial variations in Chl-a concentration and temperature in the sea area where Lekima (Typhoon No. 9 in 2019) passed through.

2. Materials

The study site is the sea area around Typhoon Lekima path, with a range of 14°~40° N, 116°~136° E (Figure 1). The data used in this study include wind, sea surface temperature (SST) and chlorophyll-a (Chl-a). The wind, sea surface temperature (SST), Chl-a and Ekman pumping velocity (EPV) were analyzed using MATLAB R2023a software.

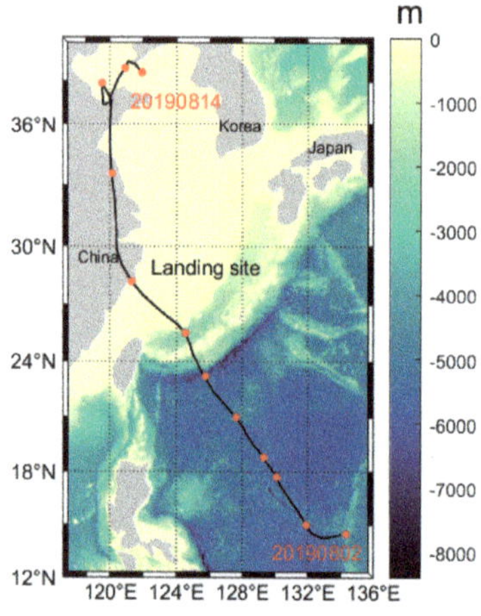

Figure 1. The route of Typhoon Lekima No. 9 in 2019.

2.1. The Overview of Typhoon Lekima No. 9 in 2019 [22,23]

On the afternoon of 30 July 2019, a tropical disturbance formed over the eastern waters of the Philippines. On the afternoon of 3 August, this developed to a tropical depression. On the afternoon of 4 August, the tropical depression officially strengthened to Lekima, the ninth typhoon of the year, and strengthened to super typhoon status at 23:00 on 7 August (Beijing time, the same below). It landed in Chengnan Town, Wenling City, Taizhou City, Zhejiang Province, China, on 10 August at 01:45, ranking as the third strongest typhoon landing in Zhejiang since the founding of the People's Republic of China (the intensity of landing was a super typhoon, and the maximum wind speed near the center was 52 m/s, level 16). After landing, the intensity gradually weakened, passing through Zhejiang and Jiangsu Province and then moving northward through the Yellow Sea. At 20:50 on 11 August, it made a second landing in the coastal area of Huangdao District, Qingdao City, Shandong Province, and then dissipated in the Bohai Sea, with a life span of 9 days. The specific path is shown in Figure 1. In Figure 1, the black line from the southeast to the northwest is the track of Typhoon Lekima.

The black line represents Lekima's path, and the red dot represents the typhoon center.

From 08:00 on 8 August to 08:00 on 11 August, the average rainfall in Zhejiang Province was 165 mm. There are mainly two heavy rainfall areas: one is located near the typhoon center, and the other is located in the area from the north of Taizhou to the south of Ningbo. In addition, there are some heavy rainfalls in the northwest of Zhejiang Province. In Taizhou, the typhoon landed in 9 counties (cities, districts); there were 5 areas with rainfall exceeding 300 mm; and the maximum rainfall was 370.9 mm in Huangyan District. The maximum precipitation measured by a single station was Kuocangshan station in Linhai City, and the accumulated precipitation during the process was 834.3 mm. The maximum hourly rainfall intensity was 896.6 mm in Ximen island of Yueqing, Wenzhou City.

2.2. Wind

Wind data were acquired from the dataset of the Navy Global Environmental Model (NAVGEM), which is a global numerical weather prediction computer model run by NOAA. In mid-February 2013, it replaced NOGAPS as the primary model for FNMOC weather model weather sites. NOGAPS is the global weather forecast model from the "Fleet Numerical Meteorology and Oceanography Center" (USA). It is located 10 m above the sea surface (spatial resolution of 1 degree) and is provided by the Navy Fleet Numerical Meteorology and Oceanography Center (FNMOC). This study period is from 3 August to 12 August 2019 (https://coastwatch.pfeg.noaa.gov/erddap/griddap/erdNavgem1D10mWind_LonPM180.html, accessed on 22 June 2024).

2.3. Sea Surface Temperature (SST)

The study period is from 1 July to 31 August 2019. The SST used in this study is from the dataset of Advanced Very High-Resolution Radiometer (AVHRR) Pathfinder Version 5.3 Level 3-Collated (L3C) SST, global, 0.0417°, daytime (1 Day Composite), provided by the National Centers for Environmental Information, USA. This is a retrospectively processed skin-level SST dataset available from 1981 to the present for climatological application [24]. This dataset comprises twice-daily (day/night) data, which are generated with measurements combined from a single AVHRR instrument (at a time) into a space-time grid. Thus, a dataset for time is a combination of multiple channels/scenes grouped together (https://coastwatch.pfeg.noaa.gov/erddap/griddap/nceiPH53sstd1day.html, accessed on 22 June 2024).

2.4. Chl-a

The data used for this study are from the typhoon period. The daily Chl-a concentration was obtained from the National Oceanic and Atmospheric Administration (NOAA) Coast Watch. The NOAA Multi-Sensor Level-1 to Level-2 (MSL12) Ocean Color, with quality science, multi-sensor global gap-filled analysis, includes Chl-a products. The data gaps were generated by using the data interpolating empirical orthogonal function (DINEOF) method [25]. The data currently come from 3 instruments: the Visible Infrared Imaging Radiometer Suite (VIIRS) sensor (aboard the Suomi National Polar-Orbiting Partnership Satellite), the VIIRS on the NOAA-20 satellite and the Ocean and Land Color Instrument (OLCI). The level 3 Chl-a (mass concentration) data over the sea surface used for this study have a spatial resolution of 4 km and are obtained daily from the National Oceanic and Atmospheric Administration (NOAA) Coast Watch/Ocean Watch. The data obtained for this study period are from August 2019.

Light reflected from or transmitted through a volume of water can be measured on a large scale to determine the average or mean influence of particles interacting with the light field; the approaches to characterizing the individual particle's size and optical characteristics show heterogeneity that is not addressed by large-scale or bulk measurements; it is typically not appropriate to determine the average properties of a whole volume solely on the average properties of individual particles or elements in the medium because most biological optical processes are non-linear [26]. The model needs to take this heterogeneity into account in a rigorously mathematical manner to link observations taken at different scales [27]. Logarithmic transformations are frequently employed when working with data that are obviously not regularly distributed, particularly in the biological and optical sciences; for bio-optical variability over a range of space-time scales, the lognormal model is helpful; these kinds of data can be analyzed using a rigorous statistical framework that the lognormal model offers; since log-converted bio-optical data tend to be normally distributed in many circumstances, the technique was based on the lognormal distribution [28]. Because the value range of sea surface Chl-a is large, it is not convenient for cartographic analysis; therefore, it is standardized (its logarithm is calculated) to achieve the compression range and facilitate the analysis of its temporal and spatial changes (https://coastwatch.star.nesdis.noaa.gov/pub/socd1/mecb/coastwatch/viirs/science/L3/global/chlora/dineof/2019/, accessed on 22 June 2024).

2.5. *Ekman Pumping Velocity (EPV)*

Ekman pumping refers to the convergence (divergence) of air at the top of the Ekman layer due to turbulent friction. The Ekman pumping effect enhances the exchange of momentum, heat and water vapor between the free atmosphere and the boundary layer [29]. The Ekman pumping velocity (EPV) is calculated from the wind stress and Coriolis parameters by using Climate Data Toolbox for MATLAB [1,30–32]. The formula is shown in Equation (1) [2]:

$$W_{EK} = (\text{curl}\ \tau)\ (\rho\ f) \tag{1}$$

In Formula (1), W_{EK} represents the EPV (m/s); τ is the surface wind stress (N); ρ is the seawater density (1025 kg/m^3); f is the Coriolis parameter (=2Ω sin θ, Ω and θ are the angular velocity and latitude of the earth, respectively). The calculation method of surface wind stress is shown in Equation (2) [1]:

$$T = [\tau_x, \tau_y] = \rho_a\ C_D\ V\ [u, v] \tag{2}$$

In Formula (2), τ_x and τ_y represent the east–west and north–south components of wind stress (N); ρ_a is the density of sea surface air; C_D is the drag coefficient; V [u, v] is the wind speed (m/s); and u and v are the east–west and north–south components of wind speed.

3. Results

3.1. *Wind Variation during the Movement of Typhoon Lekima*

Lekima formed in the northwest Pacific Ocean and dissipated over the Bohai Sea [23]. It had the characteristics of "strong landing intensity, long land retention time and wide impact area", which caused serious casualties and economic losses to many coastal provinces of China [22].

The wind is measured at 10 m above sea level with a spatial resolution of 1 degree and a temporal resolution of 6 h; the period is between 3 and 12 August 2019. The vector average analysis of the daily wind speed and direction in the passing area during the typhoon is carried out by using MATLAB software (see Figures 2 and 3). Figure 2 illustrates the variation in wind speed and direction in the area affected by the typhoon. The blue line indicates the typhoon path; the red dots represent the typhoon center; and the vectors (in black) represent wind direction and speed. Figure 3 shows the spatial distribution of wind speed; the transition from blue to red indicates that the wind speed increases from low to high. From 3 to 12 August, the typhoons experienced formation, strengthening and extinction. The wind speed is high in the area around the typhoon. In Figure 3, yellow or red-orange is the location of the typhoon center (from 3 to 12 August; some areas in the southeast from 10 to 12 August, which is the strong wind area affected by Typhoon Krosa No. 10, 2019). As the typhoon approaches closer to the center, the wind speed will be higher, and the wind direction will be counterclockwise.

3.2. SST Changes during Typhoon Lekima's Passage

In a large-scale air–sea interaction, the effect of the atmosphere on the ocean is mainly dynamic, while the effect of the ocean on the atmosphere is mainly thermal. SST represents the comprehensive results of ocean thermal and dynamic processes and the interaction between the ocean and the atmosphere [33]. SST is an important parameter, reflecting the heat, dynamic and water vapor exchange between the sea and air, and it is an important factor affecting climate change, marine nutrient concentration and primary productivity. In recent years, tropical storms have become more and more intense, which is closely related to the rise in ocean temperature [34].

Typhoon Lekima, which formed at 17:00 (Beijing time) on 4 August, became a strong tropical storm at 02:00 on 6 August, evolved into a typhoon at 05:00 on 7 August, strengthened into a strong typhoon at 17:00 and became a super typhoon at 23:00 on 7 August, and landfall occurred in Zhejiang Province at 01:00 on 10 August. In the process of typhoon formation and evolution, the regional temperature changed a lot. In the area where the typhoon passes, the SST decreases to varying degrees. The changes in the ocean and atmosphere are coupled as a whole, not isolated from each other. MATLAB software was used to analyze the spatial changes in SST during Lekima (from 3 to 12 August) and the changes in daily mean SST in the study area in August. The results are shown in Figures 4 and 5.

Figure 4 shows the spatial (14°~40° N, 116°~136° E) variation in SST before and after Lekima. In Figure 4, the white solid line is the path that the typhoon had passed; the red dotted line is the path that the typhoon will pass; and the black dot is the typhoon center. The SST change in the typhoon-affected area is obvious. In Figure 4, the temperature in the area near to the typhoon's path dropped significantly from 6 to 12 August. Before Lekima (3 August), the SST in most parts of the north (32°~40° N) was lower than 28.3 °C; however, in most parts of the middle and south (14°~32° N), the SST was higher than this. On 4 August, the SST in most of the area was in the range of 28.8~30.0 °C (14°~40° N, 116°~136° E); when Lekima passed through, the SST dropped significantly after 5 August, especially in the track area of Lekima. From 6 August, a low SST zone (around 28 °C) appeared on both sides of the typhoon track, and then, the low temperature zone gradually expanded. SST gradually increased until 15 August; however, it was still lower than before the typhoon had passed through. Another phenomenon is that as the typhoon moves forward, the SST in the early track area gradually rises.

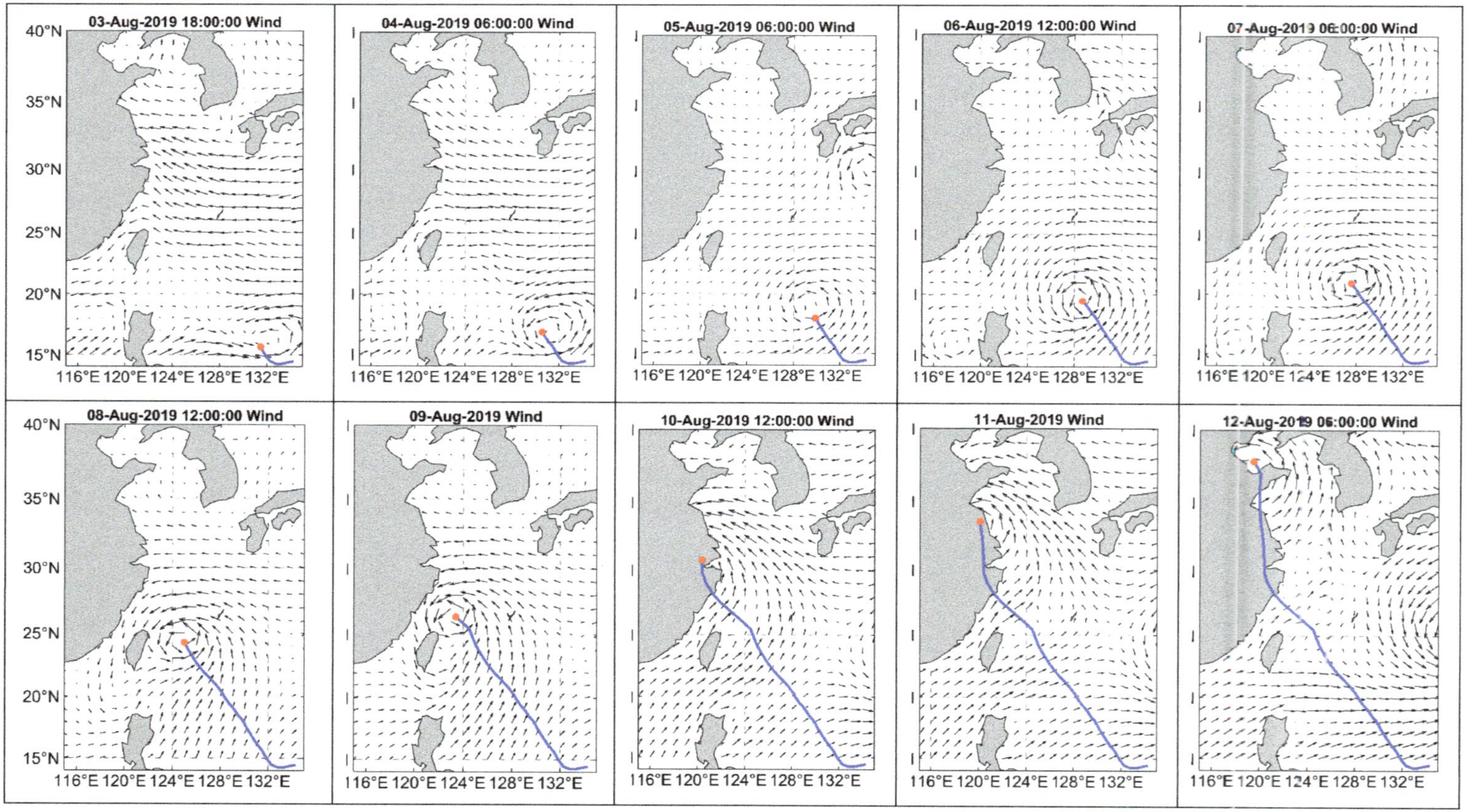

Figure 2. Wind direction and track of Typhoon Lekima. The blue line is the typhoon path; the red dot is the typhoon center; and the black arrow is the wind direction and speed (long arrows have higher wind speed and vice versa).

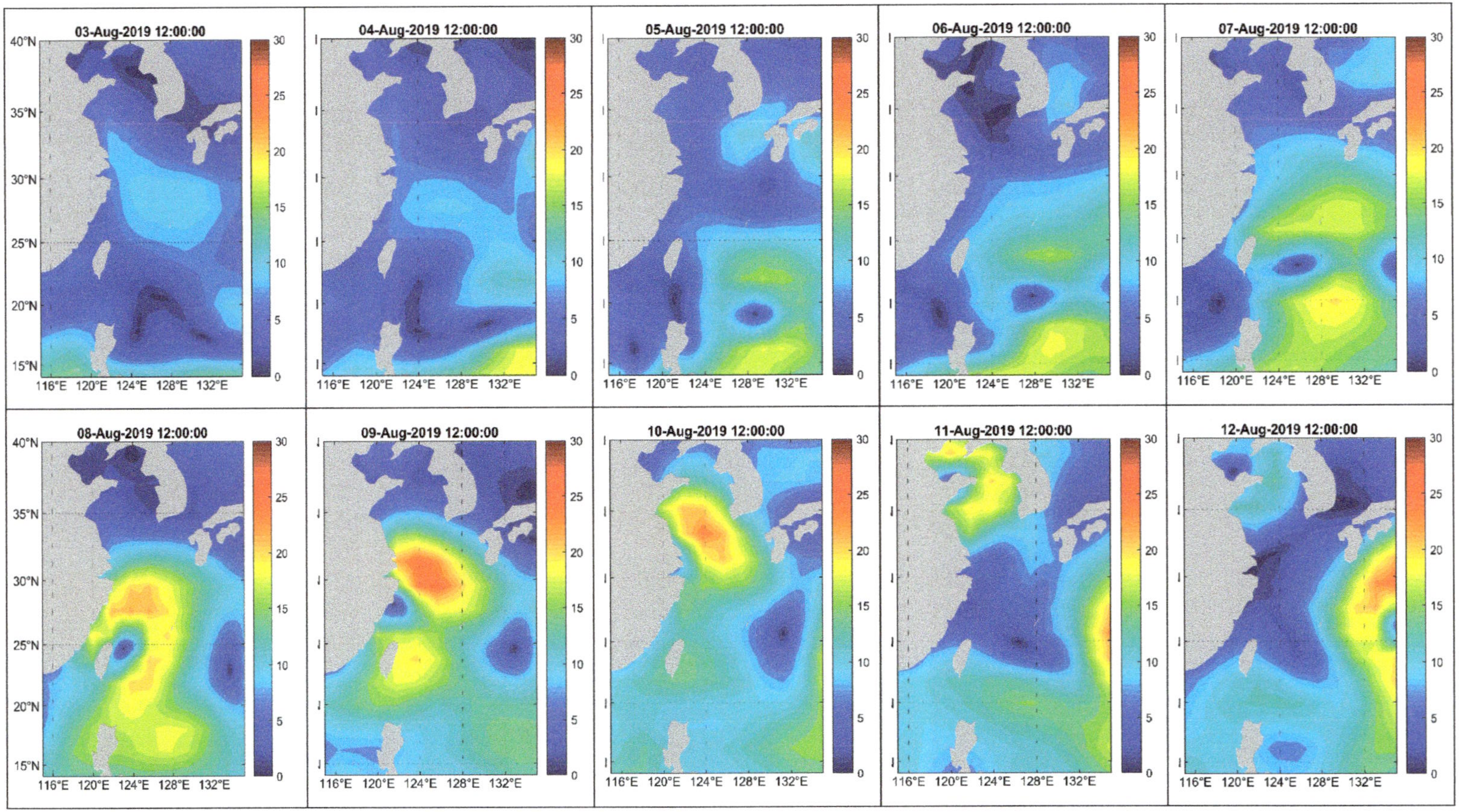

Figure 3. Wind speed during Typhoon Lekima's passage. Unit: m/s. The transition from blue to red indicates an increase in wind speed from low to high.

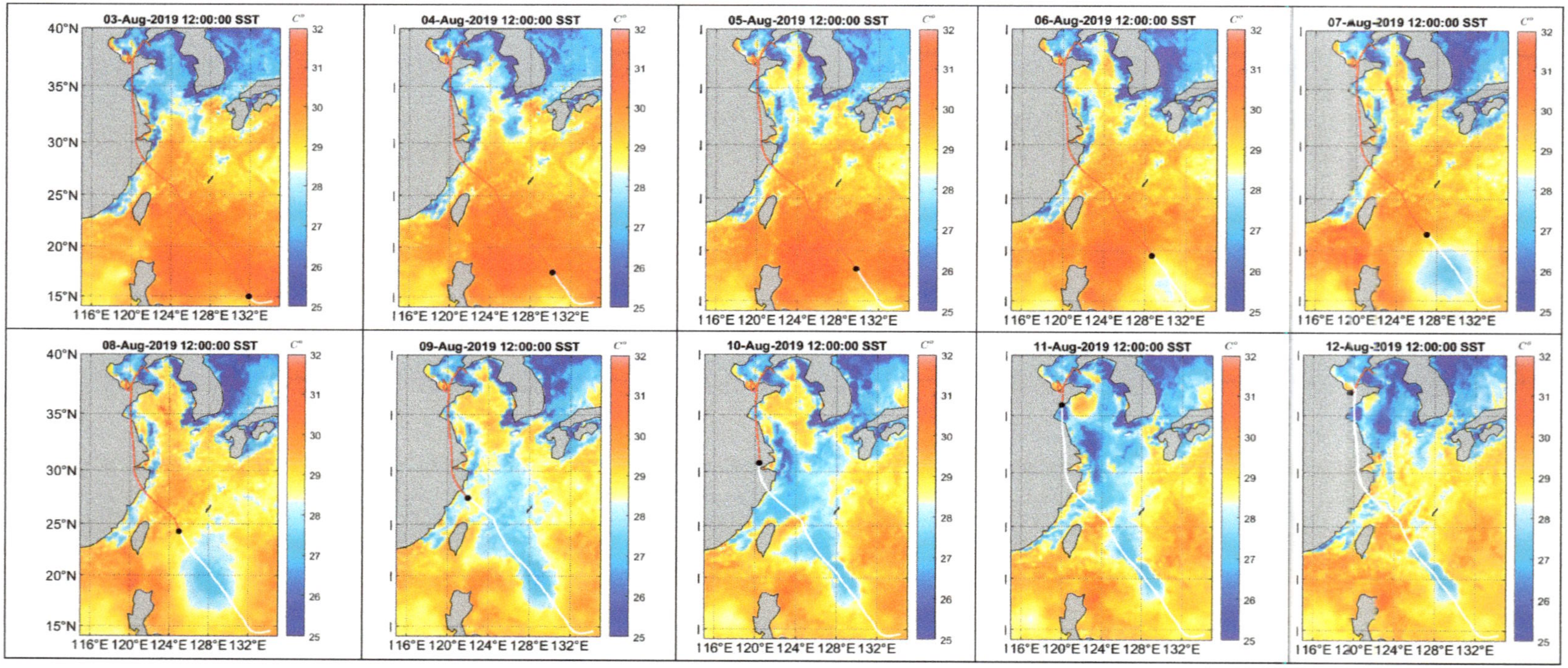

Figure 4. The temperature changes around the track of the typhoon before and after its passage. The line from the southeast to the northwest is the path of Typhoon Lekima. White represents the track that the typhoon had passed, and red represents the track that the typhoon will pass. The black dot is the typhoon center.

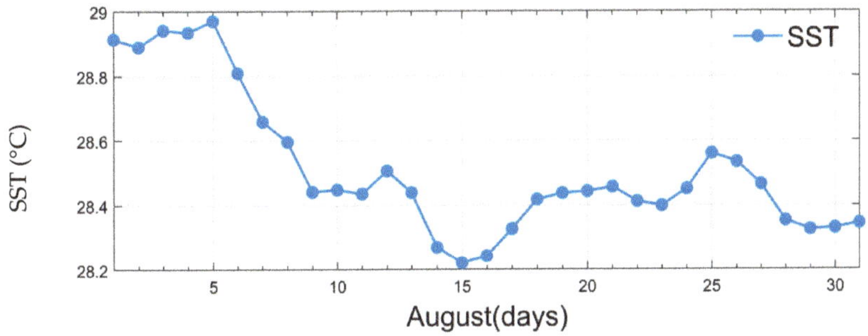

Figure 5. Changes in daily mean SST over the passage of Lekima from 1 to 31 August 2019.

Figure 5 shows the daily mean SST in the area (14°~40° N, 116°~136° E), which was affected by Lekima in August 2019; the mean monthly SST was 28.5 °C in August. From 1 to 5 August, the daily mean SST was higher than 28.89 °C, and the maximum was 28.97 °C on 5 August. With the strengthening of Lekima, the daily mean SST decreased from 28.97 °C (5 August) to 28.22 °C (15 August). After 15 August, the SST increased slightly. The SST drop is due to the increase in wind and seawater mixing when the typhoon passes, and strengthened wind can cause the deepening of the mixing layer [35]. The cyclonic wind stress on the sea surface triggered will cause the upwelling of the cold seawater in the lower layer within a radius of hundreds of kilometers, resulting in the uplift of the thermocline. Such changes in the thermocline will last for several inertial shock cycles after typhoons [21,22]. The daily mean SST in the study area dropped 0.75 °C during Lekima's passage (see Figure 5). The typhoon not only affects the SST but also the Chl-a in the seawater.

3.3. EPV Changes during Typhoon Lekima's Passage

Based on the hourly wind speed at 10 m sea surface, the EPV and its spatial distribution before and after the typhoon were calculated by MATLAB according to Formulae (1) and (2). The EPV spatial variations (14°~40° N, 116°~136° E) during Typhoon Lekima are given in Figure 6. A positive EPV value indicates an upward flow, while a negative EPV value indicates a downward flow (see Figure 6). The EPV over the ocean can clearly be seen in the figures around the typhoon. However, when the typhoon reaches the coast, the EPV is clearly seen along the coast on the right side of the typhoon, which is due to typhoon winds. When the typhoon made the landfall, the wind pattern changed, as well as the EPV, on 10 August; however the EPV disappeared on 12 August.

When a typhoon passes through, a strong wind stress acts on the upper ocean, causing strong vertical mixing in the upper ocean to deepen the depth of the mixing layer. Upwelling and downwelling caused by the convergence and divergence of mixed layer waters are the main dynamic processes of the ocean. In this paper, the EPV during the transit of Typhoon Lekima is calculated according to Formulae (1) and (2) based on wind field data acquired 10 m from the sea surface. The EPV is proportional to the wind speed in Formulae (1) and (2). Table 1 shows the maximum, average and minimum values of EPV during typhoon transit. The maximum value (positive value) indicates the upward direction of Ekman suction; the minimum value (negative value) indicates the downward direction of Ekman suction; and the mean value represents the average value of EPV within the study range (14°~40° N, 116°~136° E). On 3 August 2019, the typhoon just entered the country; the cyclone intensity increased; and the EPV increased too. The maximum, average and minimum EPV values were 1.45×10^{-4} m/s, 6.19×10^{-7} m/s and -1.48×10^{-4} m/s, respectively. Afterward, the wind speed gradually increased, and the cyclone intensified. On 8 August 2019, the cyclone intensity was the strongest, and the corresponding maximum, average and minimum values of the EPV were 7.79×10^{-4} m/s, -3.47×10^{-4} m/s and -5.31×10^{-4} m/s, respectively. On 9 August 2019, the cyclone approached the coast until landfall; the wind speed decreased; and the Ekman suction weakened. On 12 August, the cyclone disappeared, and the corresponding maximum, average and minimum EPV values were 3.43×10^{-4} m/s, 4.40×10^{-7} m/s and -1.51×10^{-4} m/s, respectively.

3.4. The Spatial Variation in Chl-a during Lekima's Passage

MODIS inversion of Chl-a's algorithm is based on the absorption and reflection effects of Chl-a in seawater. Chl-a in water shows a strong absorption effect at 440–490 nm (blue light) and high reflectivity peak at 500–550 nm (green light) [36]. Because green light can reflect the Chl-a of phytoplankton in seawater, the reflectivity of green light band will increase with the increase in Chl-a concentration [37]. The inversion formula of Chl-a concentration is based on the correlation between reflectance and Chl-a concentration in the two bands [38,39]. The formula is based on the logarithmic form of the regression equation, such as Formula 1. This formula is also applicable to Chl retrieval from the coastal zone color scanner (CZCS), ocean color and temperature scanner (OCTS) and MODIS water color remote-sensing data [40]. The coefficients in Formula (1) correspond to different satellites, and the coefficient values are calculated and determined according to the Chl-a concentration measured at fixed points and the reflectance of blue-green bands.

$$log_{10}(chlor\_a) = a_0 + \sum_{i=1}^{4} a_i log_{10}\left(\frac{R_{rs}(\lambda_{blue})}{R_{rs}(\lambda_{green})}\right)^i \quad (3)$$

In Formula (3), $R_{rs}(\lambda_{blue})$ is the higher value of reflectivity in the blue band—that is, the higher value in MODIS Band 9 (λ = 443 nm) or Band 10 (λ = 488 nm); $R_{rs}(\lambda_{green})$ is the reflectivity of the green band—that is, MODIS B and 11 (λ = 550 nm); a_i is the coefficient. $a_0 - a_4$ are 0.242, −2.7423, 1.8017, 0.0015, −1.2280, respectively; these were determined by the NASA Bio-Optical Marine Algorithm Dataset [39].

In most cases, the logarithmic transformation of bio-optical data tends toward normal distribution. The logarithm of Chl-a was normally distributed; Chl-a was converted by log_{10} for drawing and analysis [39]. In this study, the logarithmic transformation of Chl-a concentration was carried out in MATLAB software, as shown in Figure 7 [27,39]. The common logarithm is a monotonically increasing function. Figure 7 is the Chl-a concentration (common logarithm) in the sea surface during Lekima's passage (from 4 to 11 August). The distribution pattern of Chl-a (Figure 7) is different from that of SST. The high values of Chl-a were greater than 0.5 log_{10} mg/m^3, which were found near the coast, and the lower values were found far offshore. The high value area is 27°~40° N, 120°~128° E, especially in the estuary area of the Yellow River and the Yangtze River (38° N, 119° E; 31° N, 121° E). The red dot in Figure 7 is the typhoon center, and the line is the moving track (the white solid line is the path that the typhoon had passed, and the red dotted line is the path that the typhoon will pass). It can be seen that Chl-a in the nearshore changes significantly before and after the typhoon's passage. Before the typhoon's passage, the Chl-a concentration is low in the nearshore, while it increases significantly after the typhoon passes. On 10 August, Lekima landed in Chengnan Town, Wenling, Taizhou city. On that day, the Chl-a concentration in this area increased significantly compared with the previous value (the value was in the range of 0.5–1.5 log_{10} mg/m^3).

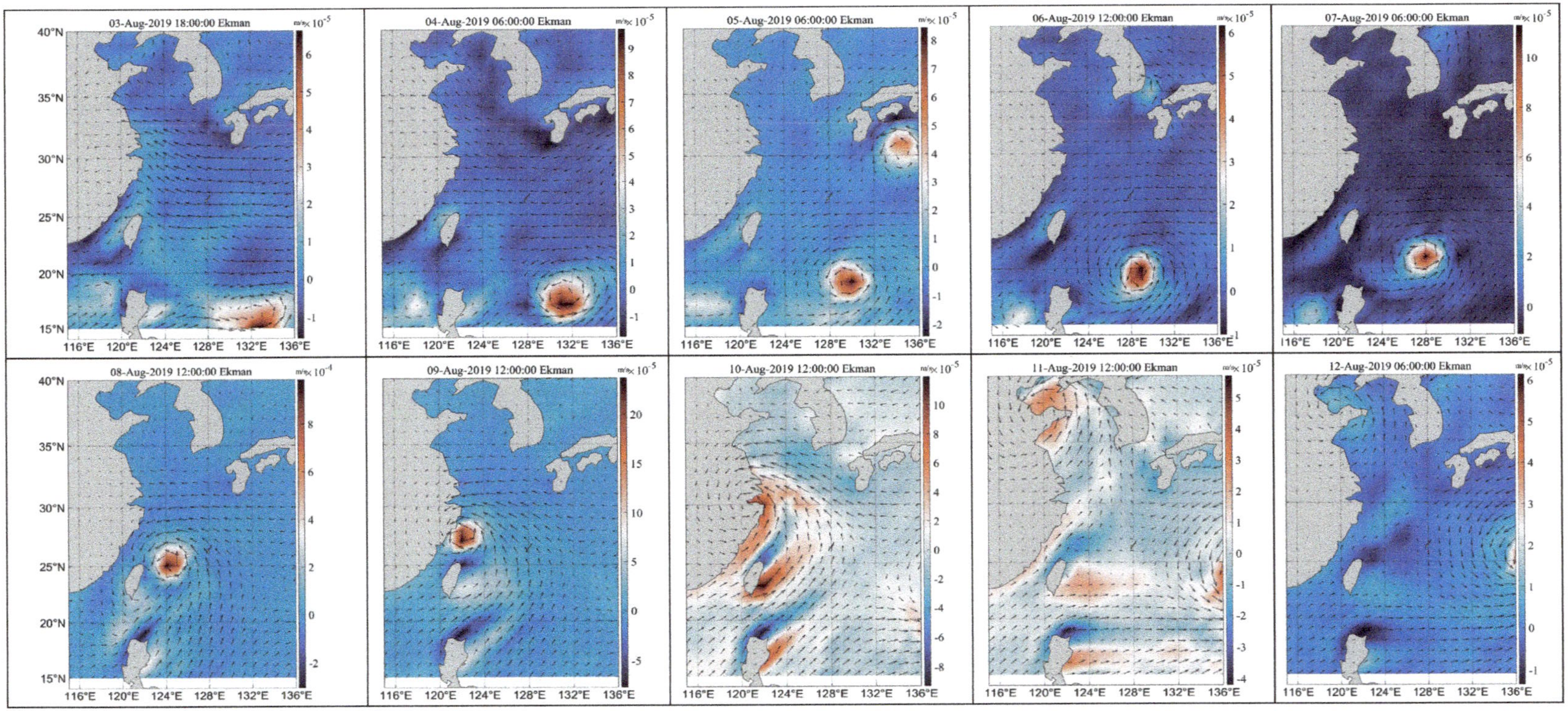

Figure 6. Spatial variation in EPV during Lekima's passage (2019). The direction of the arrow represents the wind direction, and the length represents the wind speed. The background color ranges from blue to red, indicating a gradual increase in wind speed.

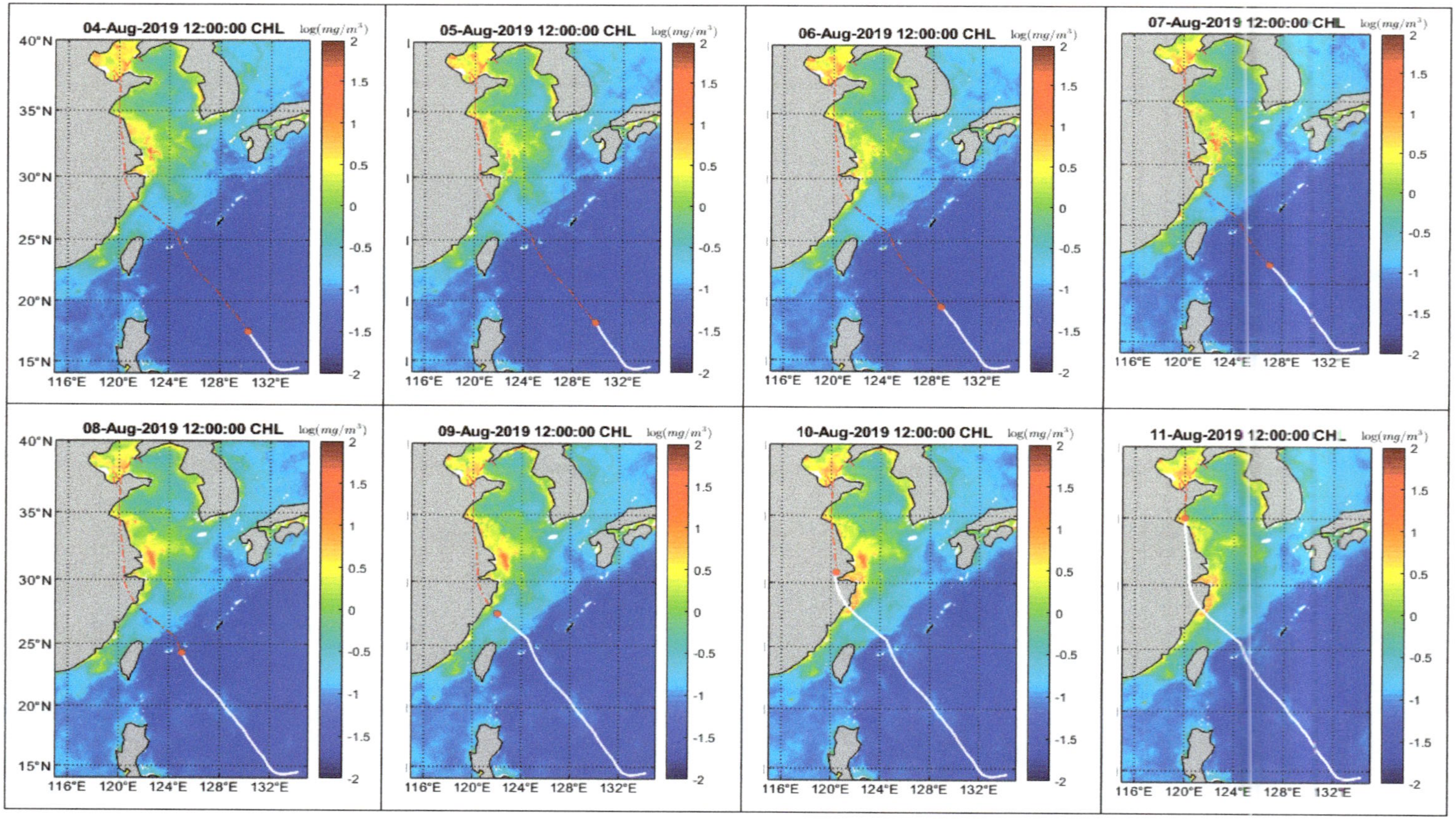

Figure 7. Spatial and temporal distribution of sea surface Chl-a concentration during Lekima's passage. Chl-a was converted to a logarithmic scale. Chl-a (log_{10} (mg/m^3)). The line from the southeast to the northwest is the path of Typhoon Lekima. White represents the track that the typhoon had passed, and red represents the track that the typhoon will pass. The red dot is the typhoon center.

Table 1. Variation in EPV during Lekima's passage (2019).

Time	Maximum (m/s)	Minimum (m/s)	Mean (m/s)
3 August 2019 18:00	1.4463×10^{-4}	-1.4769×10^{-4}	6.1895×10^{-7}
4 August 2019 06:00	1.9781×10^{-4}	-1.1461×10^{-4}	9.9883×10^{-7}
5 August 2019 06:00	1.9332×10^{-5}	-1.8586×10^{-5}	1.0662×10^{-7}
6 August 2019 12:00	5.9190×10^{-5}	-4.6737×10^{-5}	3.3041×10^{-7}
7 August 2019 06:00	1.1307×10^{-4}	-1.2251×10^{-4}	2.0743×10^{-7}
8 August 2019 12:00	7.7927×10^{-4}	-5.3071×10^{-4}	-3.4666×10^{-6}
9 August 2019 12:00	2.3698×10^{-4}	-1.3785×10^{-4}	-1.5624×10^{-6}
10 August 2019 12:00	1.5550×10^{-4}	-1.5578×10^{-4}	-6.4482×10^{-8}
11 August 2019 18:00	2.2333×10^{-4}	-6.3961×10^{-5}	1.3612×10^{-6}
12 August 2019 18:00	3.4288×10^{-4}	-1.5077×10^{-4}	4.3975×10^{-7}

Maximum, maximum upward EPV in the area. Minimum, maximum downward EPV in the area. Mean, the average EPV in the area.

4. Discussion

Typhoons can affect the ocean's dynamic factors and subsequently cause changes in the ocean's parameters. The decrease in SST and the rise in Chl-a were the main characteristics of the typhoon's passage over the area (Figures 2–4 and 7). This is mainly because the mixing and upwelling caused by typhoons weaken the stratification of the water column [41].

4.1. Response of SST to Typhoon Lekima

After Lekima had passed, there was an obvious low-temperature area around the typhoon's path. A warm sea surface can provide favorable conditions for the upward transportation of heat and water vapor. Under the influence of the typhoon, the temperature in the region (14°~40° N, 116°~136° E) where the typhoon passed decreased by a maximum of 5 °C [42,43]. In the formation and infancy of Lekima (1–5 August), the daily average SST in the study area (14°~40° N, 116°~136° E) was over 28.9 °C; the lowest was 28.2 °C on 15 August, which decreased by 0.7 °C. During the period from Lekima entering the East China Sea to the offshore, the SST was maintained at 28~30 °C, which means that the lower atmosphere over the East China Sea had abundant warm and humid air energy and vigorous sea surface evaporation. This not only ensured the maintenance of the intensity of Lekima before landing but also created favorable conditions for the generation of heavy precipitation. The changes in the ocean and atmosphere during typhoons are a coupled whole rather than isolated from each other. Typhoons receive energy from the ocean, intensify and then act on the ocean, forming large areas of low SST. As the ocean cools, it provides less energy, and the typhoons weaken.

Based on the daily average SST in July and August 2019 in the study area, the spatial distribution of the monthly average SST was calculated. In order to better understand the spatial variation in SST during the typhoon's passage, SST during the typhoon process was compared with the average SST in July and August 2019. The anomalies of SST on each day during Lekima's passage are shown in Figure 8. The figure indicates SST variation in the affected region compared with the mean SST in July and August, and the range of temperature change is approximately 0~5 °C (increase and decrease). Before the typhoon (2–4 August), the SST in the north was 1~2 °C higher than that in the center, south and also on both sides of the typhoon track. As the typhoon progressed (starting on 5 August), the SST in the typhoon path decreased significantly. In the typhoon track area, the maximum cooling range occurred during the arrival of the typhoon and one day later, and SST dropped approximately 1~3 °C; then, SST gradually increased near the typhoon track. From 7 August, there was widespread cooling in the south, with SST dropping approximately 2 °C in the area where the typhoon passed. However, in most northern areas where the typhoon did not pass (such as on 8 August), the SST increased by more than 2 °C. As the center of the typhoon approached (8–14 August), the SST in the northern region dropped from south to north. After Lekima's passage, the SST gradually increased but still kept the temperature lower than before. This shows that SST did not return to the initial state during this period, and the increase in SST took time.

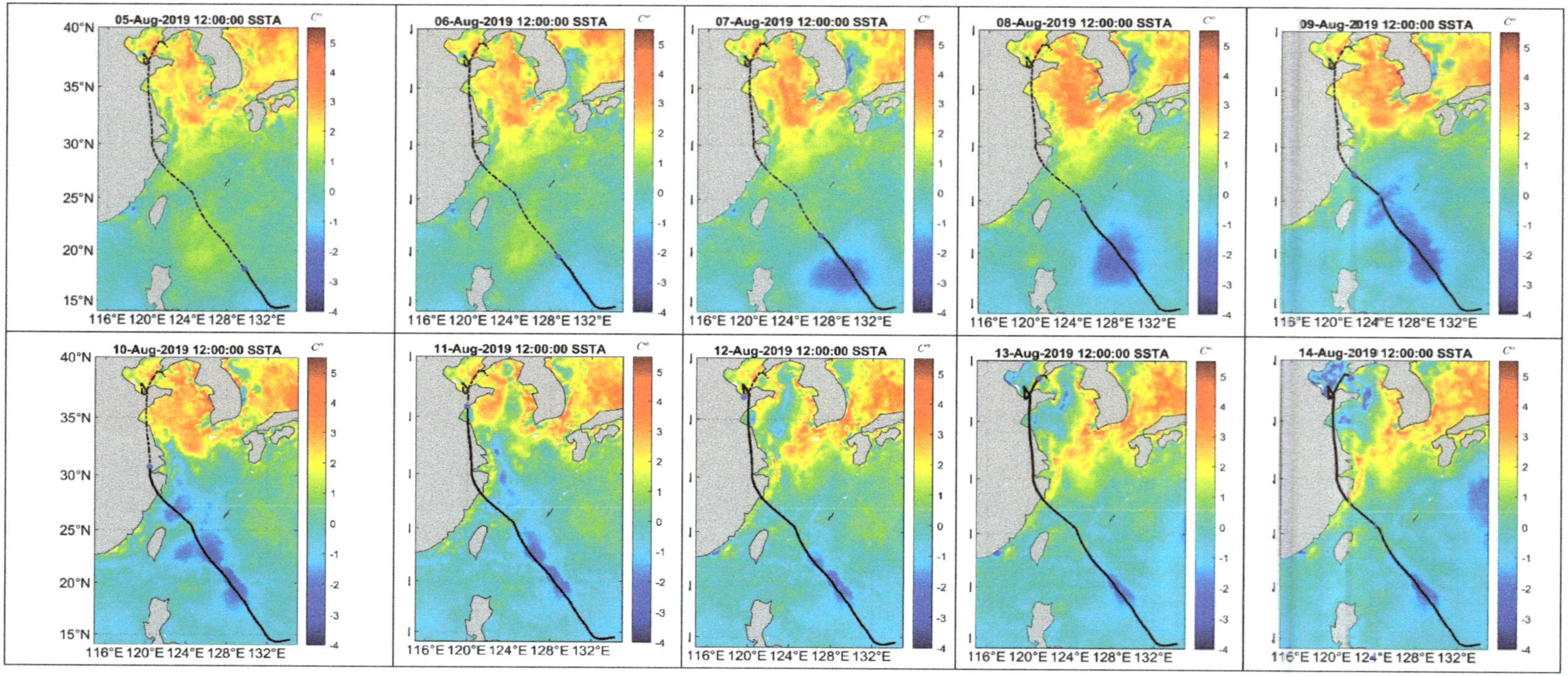

Figure 8. Response of SST to Typhoon Lekima. The line from the southeast to the northwest is the path of Typhoon Lekima. Solid line represents the track that the typhoon had passed, and dotted line represents the track that the typhoon will pass. The blue dot is the typhoon center.

The location of the main low-temperature water region caused by the typhoon basically corresponds to the area with the strongest wind speed (combining Figures 2–4 and 8). The higher the wind speed is, the greater the wind stress will be, and the stronger the mixing that will be generated in the upper ocean. Due to the diffusion effect of ocean currents on the low-temperature zone, this area will be slightly larger. Upwelling, vertical mixing, suction, rainfall, low solar radiation and high latent heat flux are the main causes of sea surface cooling caused by typhoons [4,44]; the stronger the typhoon intensity, the stronger the suction of the ocean [45].

4.2. Response of SST to Typhoon Lekima at Zhejiang Coastal Region

Figure 9 shows the daily mean SST in the area, which was affected by Lekima in August 2019, at Zhejiang coast (115~135° E, 15~35° N). On 1 August, the daily mean SST was 28.9 °C, and the maximum was 29.5 °C on 17 August. With the strengthening of Lekima, the mean SST decreased from 28.97 °C (6 August) to 13.7 °C (9 August). After 6 August, the SST increased. The SST drop is due to the increase in wind and seawater mixing when the typhoon passes. The cyclonic wind stress on the sea surface triggered will cause the upwelling of the cold seawater in the lower layer within a radius of hundreds of kilometers, resulting in the uplift of the thermocline. Such thermocline variations will last for several days after typhoons [21,22]. The average SST in the Zhejiang coastal area dropped 15.2 °C during Lekima's passage over the coastal area (see Figure 9) due to strong upwelling and rainfall, attaining the maximum value on 17 August. The typhoon not only affects SST during passage; it also affects the Chl-a in the seawater after passage.

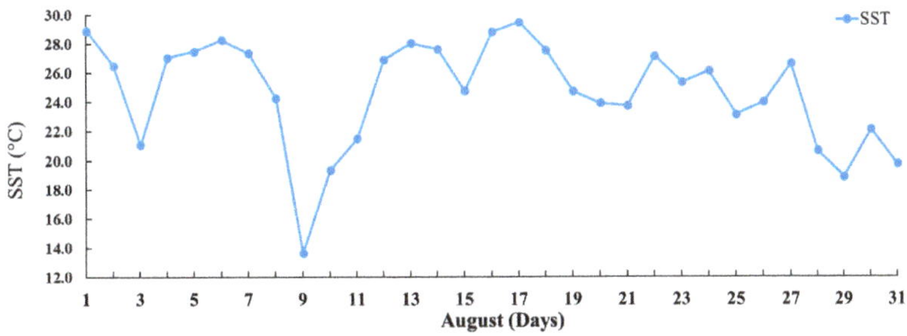

Figure 9. Changes in daily mean SST over the passage of Lekima in Zhejiang coastal area from 1 to 31 August 2019.

4.3. Response of Chl-a to Typhoon Lekima

The variation in Chl-a is influenced by the SST, wind speed and ocean current, etc., and its spatial distribution reflects the large-scale vertical transport of nutrients driven by these factors [46–48]. The upwelling, suction and entrainment caused by typhoons (please refer to Figure 6, where the EPV is located along the coast) bring the nutrients in the subsurface layer to the upper layer and promote the growth of phytoplankton and Chl-a, which has an important impact on the primary productivity and carbon cycle in the oligotrophic sea area [49]. When the EPV is located along the coast, there is upwelling occurring near the coastal areas, and further, the Chl-a increases due to the surface water moving away from the coast and subsurface nutrient water appearing on the surface. The typhoon caused a significant increase in the concentration of Chl-a on the sea surface in the cold-water area, which was related to the bringing of nutrients and biological pigments to the surface and would last for 2~3 weeks before returning to the level before the typhoon [50]. Typhoon enhanced Chl-a and affected the ecosystem mainly by changing nutrient content through vertical mixing and transportation. In addition, typhoon-related rainfall will increase the river flow into the sea and then increase the intensity of land nutrients as input [50]. However, the response of Chl-a to typhoons varies greatly, which depends on water depth, typhoon speed and intensity [51].

Chl-a variation is obvious in the coastal area of Typhoon Lekima's passage (Figure 10, range: 14°–40° N, 116°–136° E). Chl-a in the nearshore is higher than that in the offshore. Chl-a is low before Lekima's passage, while it is high after the typhoon's passage over the coastal area. After the typhoon's passage, there was an obvious increase in Chl-a around the typhoon's path, which increased rapidly in a short time, and the location was consistent with the low-temperature area. Figure 10 indicates the variations in average Chl-a concentration and mean SST over the study area in August 2019. Following Lekima's passage, the mean SST oscillated, while the Chl-a concentration tracked fluctuations in the SST. The change in SST was negatively correlated with the change in Chl- a concentration before and during the typhoon (from 1 to 14 August); however, the Chl-a concentration increased rapidly from 15 August (see Figure 10). As mentioned in the previous work of Fu et al. [52], the Chl-a concentration in coastal waters increased with a lag of 3–6 days. During Typhoon Lekima, the Chl-a increased after 5 days over the coastal areas of Zhejiang. Before the typhoon (2 August), Chl-a reached a peak value (0.48 mg/m^3) and then continued to decline. It reached a trough value (0.40 mg/m^3) on 8 August and increased over the following two days. It reached a peak value (0.45 mg/m^3) on 10 August and increased after 17 August, remaining at a higher value for the next 3 days.

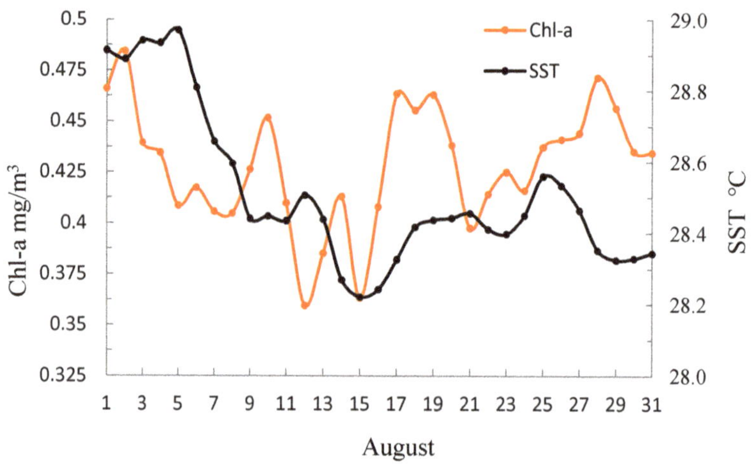

Figure 10. Chl-a concentration variation with SST during Lekima's passage.

After the typhoon, the Chl-a concentration increased significantly in the early typhoon center area (especially in the nearshore; see Figure 7), which was caused by the increase in evaporation, the enhancement of mixing and the upwelling of phytoplankton in the deep layer caused by the upwelling of thermocline. Meanwhile, the surge of cold water in the lower layer brought rich nutrients to the barren upper ocean. In addition, the sunny weather after the typhoon provides favorable conditions for phytoplankton photosynthesis and reproduction, which may also be one of the reasons why the Chl-a concentration exhibited a high value over the next few days. Because the change in the thermocline will last for several inertial cycles after the typhoon, the Chl-a on the ocean surface in the central area will fluctuate with the overall increase after the typhoon. The wind force on the periphery of the typhoon is small, which leads to mixing and thermocline surge being small; therefore, the increase in Chl-a on the periphery of the typhoon is small.

In this study, the error of the Chl-a derived from VIIRS is within 30%, and most Chl-a data are more accurate, such as the median ratio between 0.982 and 1.142 and the median error between 2 and 14% [53].

5. Conclusions

In this paper, we analyzed the impact of Typhoon Lekima on the SST and sea surface Chl-a. Typhoons can affect the ocean dynamically and then cause changes in the ocean's parameters. A decrease in SST and a rise in Chl-a were the main characteristics caused by the typhoon in the study area.

The main reason is that the mixing and upwelling caused by typhoons can attenuate the stratification of the water column [41]. Therefore, as cold and highly nutritious subsurface water rises to the surface, the SST will decrease, and Chl-a will increase [54]. When a typhoon passes by, the surface wind increases, and the mixing intensifies; therefore, it usually causes a deepening of the mixing layer. The cyclonic wind stress on the sea surface caused by typhoons can lead to cold water in the lower layer rising within a radius of hundreds of kilometers, resulting in the uplift of the thermocline [4]. The change in the thermocline will last for several inertial oscillation cycles after the typhoon's passage. After the typhoon's passage, there was a cooling process in the ocean mixing layer, and the temperature of the surface mixing layer decreased by 1 °C on average [55–57].

Typhoon Lekima increased the concentration of Chl-a on the sea surface, which was related to the nutrients and biological pigments being brought to the surface, and this would last for a period of time before returning to normal before the typhoon's passage. The upwelling, entrainment and suction caused by the typhoon bring the phytoplankton in the subsurface and the nutrients in the lower layer to the upper layer, which promotes the growth of phytoplankton and the increase in Chl-a concentration in the sea surface. Heavy rainfall, increased evaporation, enhanced mixing layer and the thermocline surge induced by the typhoon all contributed to the increase in Chl-a to varying degrees. Rainfall caused by typhoons will increase the river flow and then increase the intensity of nutrient input on land [50], which will also affect the distribution of chlorophyll near the shore. When Lekima passed through, the heavy rainfall caused land runoff and offshore flow, which carried a large amount of suspended matter, phytoplankton and dissolved organic matter into the sea, forming a higher chlorophyll area in the nearshore. This is also the reason for an increase in Chl-a concentration in the nearshore after a typhoon. Areas near the typhoon track illustrate obvious changes after the typhoon passes; the temperature decreases, and the nutrients and Chl-a increase. The response of Chl-a generally occurs later than that of the SST because the growth of phytoplankton takes several days.

Author Contributions: Conceptualization, B.G., Y.S. and V.S.M.; methodology, Y.S. and B.G.; software, Y.N., J.W., Z.J., Y.C. and M.Y.; validation, B.G. and V.S.M.; resources, V.S.M., J.W., Y.N. and M.Y.; data curation, V.S.M. and Y.N.; writing—original draft preparation, Y.S. and B.G.; writing—review and editing, Y.S., B.G. and V.S.M.; project administration, B.G.; funding acquisition, B.G. All authors have read and agreed to the published version of the manuscript.

Funding: This research was funded by the Consultation and Evaluation Program of the Department of Chinese Academy of Science (2020-ZW11-A-023), the open research fund program of State Key Laboratory of Hydroscience and Engineering, Tsinghua University (sklhse-2021-B-01), Zhejiang Provincial Virtual Simulation Experiment Projects (virtual simulation experimental project for monitoring of suspended sediment in coastal waters using remote sensing), China Scholarship Council (NO. 20230833027), National College Students Innovation and Entrepreneurship Training Program (NO. 202110340039), Zhejiang University Students Science and Technology Innovation Activity Plan (New Talent Program. NO. 2022R411A003), Open Research Fund Program of Key Laboratory for Water Ecology Management and Protection in River Source Areas, Ministry of Water Resources (No. 2024slbjh02) and the National Natural Science Foundation of China (No. 52109092).

Institutional Review Board Statement: Not applicable.

Informed Consent Statement: Not applicable.

Data Availability Statement: Data are contained within the article.

Conflicts of Interest: The authors declare no conflicts of interest.

References

1. Hun, K.S.; Ju, M.I.; Shin, C.P. An increase in global trends of tropical cyclone translation speed since 1982 and its physical causes. *Environ. Res. Lett.* **2020**, *15*, 094084. [CrossRef]
2. Guan, S.; Zhao, W.; Sun, L.; Zhou, C.; Liu, Z.; Hong, X.; Zhang, Y.; Tian, J.; Hou, Y. Tropical cyclone-induced sea surface cooling over the Yellow Sea and Bohai Sea in the 2019 Pacific typhoon season. *J. Mar. Syst.* **2021**, *217*, 103509. [CrossRef]
3. Feng, S.; Li, F.; Li, S. *An Introduction to Marine Science*; Higher Education Press: Beijing, China, 1999.
4. Price, J.F. Upper ocean response to a hurricane. *J. Phys. Oceanogr.* **1981**, *11*, 153–175. [CrossRef]
5. Ioualalen, M.; Wakata, Y.; Kawahara, Y.; Gouriou, Y.; Varillon, D. Variability of the sea surface salinity (SSS) in the Western Tropical Pacific: On the ability of an OGCM to simulate the SSS, and on the sampling of an operating merchant ship SSS network. *J. Oceanogr.* **2003**, *59*, 105–111. [CrossRef]
6. Scoccimarro, E.; Fogli, P.G.; Reed, K.A.; Gualdi, S.; Masina, S.; Navarra, A. Tropical cyclone interaction with the ocean: The role of high-frequency (sub daily) coupled processes. *J. Clim.* **2017**, *30*, 145–162. [CrossRef]
7. Zedler, S.E.; Dickey, T.D.; Doney, S.C.; Price, J.F.; Yu, X.; Mellor, G.L. Analyses and simulations of the upper ocean's response to Hurricane Felix at Bermuda Test bed Mooring site: 13–23 August 1995. *J. Geophys. Res.* **2002**, *107*, 3232. [CrossRef]
8. Xiang, J. *Marine Biology*; Science Press: Beijing, China, 2003.
9. Xu, D.; Liu, Z.; Xu, X.; Xu, J. The influence of typhoon on the sea surface salinity in the warm pool of the western Pacific. *Acta Oceanol. Sin.* **2005**, *27*, 9–15.
10. Kessler, W.S. The circulation of the eastern tropical Pacific: A review. *Prog. Oceanogr.* **2006**, *69*, 181–217. [CrossRef]
11. Hu, D.; Mao, K.; Chen, X.; Zhao, Y.; Li, Y. Composing analysis on upper ocean response to typhoon in the northwest Pacific. *J. Appl. Oceanogr.* **2018**, *37*, 506–513.
12. Walker, N.; Leben, R.R.; Balasubramanian, S. Hurricane forced upwelling and chlorophyll a enhancement within cold core cyclones in the Gulf of Mexico. *Geophys. Res. Lett.* **2005**, *32*, L18610. [CrossRef]
13. Jiang, X.; Zhong, Z.; Liu, C. The effect of typhoon-induced SST cooling on typhoon intensity: The case of Typhoon Chanchu (2006). *Adv. Atmos. Sci.* **2008**, *25*, 1062–1072. [CrossRef]
14. Chen, D.; He, L.; Liu, F.; Yin, K. Effects of typhoon events on chlorophyll and carbon fixation in different regions of the East China Sea. *Estuar. Coast. Shelf Sci.* **2017**, *194*, 229–239. [CrossRef]
15. Chen, C.; Shi, P.; Mao, Q. Satellite remotely-sensed analysis of distribution characters of chlorophyll concentration in South China Sea. *J. Trop. Oceanogr.* **2001**, *20*, 66–70.
16. Shang, S.; Hong, H.; Zhang, C.; Shang, S.; Hu, J. Chl-a distribution feature of Taiwan Straits region in winter, 1998 as observed by SeaWiFS. *Mar. Sci. Bull.* **2001**, *20*, 25–29.
17. Zhao, H.; Qi, Y.; Wang, D.; Wang, W. Study on the features of chlorophyll-a derived from SeaWifs in the South China Sea. *Acta Oceanol. Sin.* **2005**, *27*, 45–52.
18. Veeranjaneyulu, C.; Deo, A.A. Study of upper ocean parameters during passage of tropical cyclones over Indian seas. *Int. J. Remote Sens.* **2019**, *40*, 4683–4723.
19. Kwon, Y.O.; Riser, S.C. *The Ocean Response to the Hurricane and Tropical Storm in North Atlantic during 1997–1999*; School Oceanography, University of Washington: Washington, DC, USA, 2003.
20. D'Asaro, E.A. The ocean boundary layer below Hurricane Dennis. *J. Phys. Oceanogr.* **2003**, *33*, 561–578. [CrossRef]
21. Lin, I.I.; Liu, W.T.; Wu, C.C.; Chiang, J.C.H.; Sui, C.H. Satellite observations of modulation of surface winds by typhoon induced upper ocean cooling. *Geophys. Res. Lett.* **2003**, *30*, 1131. [CrossRef]
22. Zhou, C.; Chen, P.; Yang, S.; Zheng, F.; Yu, H.; Tang, J.; Lu, Y.; Chen, G.; Lu, X.; Zhang, X.; et al. The impact of Typhoon Lekima (2019) on East China: A post event survey in Wenzhou City and Taizhou City. *Front. Earth Sci.* **2022**, *16*, 109–120. [CrossRef]
23. Tang, J.; Fei, J.; Yu, H. Research of super typhoon Lekima: Forecast, observation, numerical simulation and disaster survey. *Front. Earth Sci.* **2022**, *16*, 1–4. [CrossRef]
24. Saha, K.; Zhao, X.; Zhang, H.; Casey, K.S.; Zhang, D.; Baker-Yeboah, S.; Kilpatrick, K.A.; Evans, R.H.; Ryan, T.; Relph, J.M. *AVHRR Pathfinder Version 5.3 Level 3 Collated (L3C) Global 4 km Sea Surface Temperature for 1981-Present, Dataset*; NOAA National Centers for Environmental Information: Asheville, NC, USA, 2018.
25. Wang, Y.; Liu, D. Reconstruction of satellite chlorophyll-a data using a modified DINEOF method: A case study in the Bohai and Yellow seas, China. *Int. J. Remote Sens.* **2014**, *35*, 204–217. [CrossRef]
26. Campbell, J.W. The lognormal distribution as a model for bio-optical variability in the sea. *J. Geophys. Res.* **1995**, *100*, 13237–13254. [CrossRef]
27. Yentsch, C.M.; Campbell, J.W. Phytoplankton growth: Perspectives gained by flow cytometry. *J. Plankton Res.* **1991**, *13* (Suppl. S1), 83–108.
28. Liu, X.; Wang, M. Filling the gaps of missing data in the merged VIIRS SNPP/NOAA-20 ocean color product using the DINEOF method. *Remote Sens.* **2019**, *11*, 178. [CrossRef]
29. Geider, R.J.; Macintyre, H.L.; Kana, T.M. Dynamic model of phytoplankton growth and acclimation: Responses of the balanced growth rate and the chlorophyll a: Carbon ratio to light, nutrient-limitation and temperature. *Mar. Ecol. Prog.* **1997**, *148*, 187–200. [CrossRef]
30. Kessler, W.S. Mean three-dimensional circulation in the northeast tropical Pacific. *J. Phys. Oceanogr.* **2002**, *32*, 2457–2471. [CrossRef]
31. Greene, C.A.; Blankenship, D.D.; Gwyther, D.E.; Silvano, A.; van Wijk, E. Wind causes Totten Ice Shelf melt and acceleration. *Sci. Adv.* **2017**, *3*, e1701681. [CrossRef] [PubMed]

32. Greene, C.A.; Thirumalai, K.; Kearney, K.A.; Delgado, J.M.; Schwanghart, W.; Wolfenbarger, N.S.; Thyng, K.M.; Gwyther, D.E.; Gardner, A.S.; Blankenship, D.D. The climate data toolbox for MATLAB. *Geochem. Geophys. Geosyst.* **2019**, *20*, 3774–3781. [CrossRef]
33. Bao, X.; Wan, X.; Gao, G.; Wu, D. The characteristics of the seasonal variability of the sea surface temperature field in the Bohai Sea, the Huanghai Sea and the East China Sea from AVHRR data. *Acta Oceanol. Sin.* **2002**, *24*, 125–133.
34. Kawai, Y.J.; Kawamura, H. Evaluation of diurnal warming of sea surface temperature using satellite-derived marine meteorological data. *J. Oceanogr.* **2002**, *58*, 805–814. [CrossRef]
35. Price, J.F.; Sanford, T.B.; Forristall, G.Z. Forced stage response to a moving hurricane. *J. Phys. Oceanogr.* **1994**, *24*, 233–260. [CrossRef]
36. O'Reilly, J.E.; Werdell, P.J. Chlorophyll algorithms for ocean color sensors-OC4, OC5 & OC6. *Remote Sens. Environ.* **2019**, *229*, 32–47.
37. Sasaki, H.; Tanaka, A.; Iwataki, M.; Touke, Y.; Siswanto, E.; Tan, C.K.; Ishizaka, J. Optical properties of the red tide in Isahaya Bay, southwestern Japan: Influence of chlorophyll a concentration. *J. Oceanogr.* **2008**, *64*, 511–523. [CrossRef]
38. Tan, S. Radiative Properties of Euphotic Layer and Their Impacts on Satellite Remote Sensing of Ocean Primary Productivity. Doctoral Dissertation, Chinese Academy of Sciences, Beijing, China, 2007. (In Chinese).
39. Siegel, D.; Behrenfeld, M.; Maritorena, S.; McClain, C.; Antoine, D.; Bailey, S.; Bontempi, P.; Boss, E.; Dierssen, H.; Doney, S.; et al. Regional to global assessments of phytoplankton dynamics from the seaWiFS mission. *Remote Sens. Environ.* **2013**, *135*, 77–91. [CrossRef]
40. Werdell, P.J.; Franz, B.A.; Bailey, S.W.; Harding, L.W., Jr.; Feldman, G.C. Approach for the long-term spatial and temporal evaluation of ocean color satellite data products in a coastal environment. *Coast. Ocean. Remote Sens.* **2007**, *6680*, 66800G. [CrossRef]
41. Chiang, T.; Wu, C.; Oey, L. Typhoon Kai-Tak: An ocean's perfect storm. *J. Phys. Oceanogr.* **2011**, *41*, 221–233. [CrossRef]
42. Liu, S.; Zhang, L.; Zhang, X. Characteristics analysis on rapid intensification of typhoon Mujigae (1522) over the offshore area of China. *J. Meteorol. Sci.* **2017**, *37*, 487–496. (In Chinese)
43. Song, P.; Zhong, Z.; Qi, L.; Yuan, S.; Wang, X. A numerical study on the influence of abnormal local sea surface temperature on the track of tropical cyclone. *J. Meteorol. Sci.* **2017**, *37*, 737–747. (In Chinese)
44. Sanford, T.B.; Black, P.G.; Haustein, J.R.; Feeney, J.W.; Forristall, G.Z.; Price, J.F. Ocean response to a hurricane, Part I: Observations. *J. Phys. Oceanogr.* **1987**, *17*, 2065–2083. [CrossRef]
45. Li, L.; Bao, X.; Gao, G. Numerical experiment of SST response to typhoon process in Yellow Sea and Bohai Sea. *Period. Ocean. Univ. China* **2001**, *31*, 165–172.
46. McClain, C.R. A decade of satellite ocean color observations. *Annu. Rev. Mar. Sci.* **2009**, *1*, 19–42. [CrossRef]
47. Sauer, M.J.; Roesler, C.S.; Werdell, P.J.; Barnard, A. Under the hood of satellite empirical chlorophyll a algorithms: Revealing the dependencies of maximum band ratio algorithms on inherent optical properties. *Opt. Express* **2012**, *20*, 20920–20933. [CrossRef] [PubMed]
48. Yoder, J.A.; Kennelly, M.A. Seasonal and ENSO variability in global ocean phytoplankton chlorophyll derived from 4 years of SeaWiFS measurements. *Glob. Biogeochem. Cycles* **2003**, *17*, 1112. [CrossRef]
49. Sathyendranath, S.; Gouveia, A.; Shetye, S.; Ravindran, P.; Platt, T. Biological control of surface temperature in the Arabian Sea. *Nature* **1999**, *349*, 54–56. [CrossRef]
50. Chen, Y.; Tang, D. Remote sensing analysis of impact of typhoon on environment in the sea area South of Hainan Island. *Procedia Environ. Sci.* **2011**, *10 Pt B*, 1621–1629. [CrossRef]
51. Wei, M.; Lien, C.; Lin, I.; Xie, S. Tropical cyclone-induced ocean response: A comparative study of the South China Sea and Tropical Northwest Pacific. *J. Clim.* **2015**, *28*, 5952–5968. [CrossRef]
52. Fu, D.; Pan, D.; Deng, Y.; Zou, J. Quantitative investigation on typhoon impact on chlorophyll a concentration by remote sensing. *Acta Oceanol. Sin.* **2009**, *31*, 46–56.
53. Wang, M.; Liu, X.; Jiang, L.; Son, S. Visible Infrared Imaging Radiometer Suite (VIIRS) Ocean Color Products, Algorithm Theoretical Basis Document (ATBD), Version 1.0, June 2017. Available online: https://www.star.nesdis.noaa.gov/socd/mecb/color/documents/ATBD_VIIRS_OC_v1.0_June2017_f2.pdf (accessed on 22 June 2024).
54. Baibin, S.M.; Carton, J.A.; Dickey, T.D.; Wiggert, J.D. Satellite evidence of hurricane-induced phytoplankton blooms in an oceanic desert. *J. Geophys. Res.* **2004**, *109*, C03043. [CrossRef]
55. Jaimes, B.; Shay, L.K. Near-inertial wave wake of Hurricanes Katrina and Rita over mesoscale oceanic eddies. *J. Phys. Oceanogr.* **2009**, *40*, 1320–1337. [CrossRef]
56. Rao, A.D.; Babu, S.V.; Dube, S.K. Impact of a tropical cyclone on coastal upwelling process. *Nat. Hazards* **2004**, *31*, 415–435. [CrossRef]
57. Wada, A. Numerical simulation of sea surface cooling by a mixed layer model during the passage of Typhoon Rex. *J. Oceanogr.* **2005**, *61*, 41–57. [CrossRef]

atmosphere

MDPI

Article

Increased Frequency of Central Pacific El Niño Events Since 2000 Caused by Frequent Anomalous Warm Zonal Advection

Lanyu Jia and Yongqing Guo *

Marine Science and Technology College, Zhejiang Ocean University, Zhoushan 316022, China;
jialanyu@zjou.edu.cn
* Correspondence: guoyongqing@zjou.edu.cn

Abstract: Although intensive studies have documented the recent increase in the frequency of the Central Pacific (CP) El Niño events, the underlying mechanism remains unclear. This motivates us to investigate the change in the frequency of CP El Niño events. By analyzing the occurrence of the CP El Niño events between 1960 and 2022, we confirm a statistically significant increase in the frequency of CP El Niño events since 2000. Over the 40 years between 1960 and 1999, eight CP El Niño events appeared. In contrast, over the 23 years between 2000 and 2022, six CP El Niño events are seen. The significant period of the CP El Niño shortens from 4–5 years to 2–3 years. The increased frequency of CP El Niño events is closely related to more frequent warm sea surface temperature (SST) anomalies in the central equatorial Pacific (5° S–5° N, 170° W–122° W) during the CP El Niño developing phase (June to October). A heat budget analysis of the mixed layer reveals that the SST variability in the central equatorial Pacific during the developing phase is determined by zonal temperature advection. The frequent anomalous warm zonal advection drives more frequent warm SST anomalies, and finally, the increased frequency of CP El Niño events, as observed.

Keywords: Central Pacific El Niño; El Niño frequency change; zonal temperature advection; heat budget analysis

1. Introduction

El Niño–Southern Oscillation (ENSO) is the most significant interannual climate variability on the global scale [1]. As first pointed out by Bjerknes in 1969 [2], the ENSO is an air–sea coupled climate mode. Specifically, El Niño describes the oceanic side of the coupled mode [3–6], while the Southern Oscillation depicts the atmospheric side [7]. Historically, an El Niño event is identified according to the warm sea surface temperature (SST) anomalies in the Niño3 region (5° S–5° N and 150° W–90° W), which is located in the eastern equatorial Pacific [8]. This type of El Niño is now often called the Eastern Pacific (EP) El Niño [9]. The occurrence of traditional EP El Niño events strongly influences the global climate via the atmospheric bridge [10]. For example, the strong 1997/1998 EP El Niño event caused massive floods in the Yangtze River Basin in China during the 1998 summer [11]. Meanwhile, this El Niño event brought severe storms to the southern United States and California, unusually warm and dry conditions to the northern United States, resulting in USD 4 billion in losses [12].

Since 1990, intensive studies have documented another type of El Niño, characterized by the largest warm SST anomalies confined primarily to the central equatorial Pacific instead of the eastern equatorial Pacific [13–16]. As the warm SST anomalies appear mainly

 https://doi.org/10.3390/atmos16060654

in the central equatorial Pacific near the International Date Line and close to the warm pool, this emerging type is known as Central Pacific (CP) El Niño [16,17], Dateline El Niño [18], warm-pool El Niño [6], or El Niño Modoki [15]. This study refers to it as the CP El Niño, as the counterpart of the EP El Niño. The remarkable warming over the central equatorial Pacific during the CP El Niño makes its influence on weather and climate quite different from the EP El Niño [5]. For example, during the late autumn to winter of EP and CP El Niño events, South China experiences increased precipitation. However, the positive precipitation anomalies during the EP and CP El Niño events differ in intensity and spatial pattern. The positive precipitation anomalies during the CP El Niño event are weaker and located further west, mainly in the inland region of South China, compared with the EP El Niño event [19]. The CP El Niño has a stronger influence on winter temperatures in the northwestern and southeastern United States, while the EP type mainly affects the Great Lakes, Northeast, and Southwest regions [20].

The two different types of El Niño, especially the CP type, have attracted much attention. Although some continue debating whether the CP El Niño constitutes a fundamentally distinct phenomenon or shares dynamical origins with EP El Niño while exhibiting different manifestations, intensive studies contrasted the mechanisms and variabilities of the EP and CP El Niño events. The mechanism for generating the EP El Niño events has been well-established [21–24]. Four conceptual models of ENSOs have been developed, including the delayed oscillator [21,25], the recharge oscillator [22,26], the western Pacific oscillator [23,27], and the advective–reflective oscillator [24]. However, the mechanisms for the CP El Niño events remain unclear. Four main hypotheses have been put forward. The first hypothesis attributes CP El Niño occurrence to the weakening of equatorial easterly winds. Ashok et al. (2007) [15] demonstrated that, after 1979, reductions in equatorial easterlies decreased zonal SST gradients, subsequently flattening the thermocline and creating favorable conditions for CP El Niño emergence. The second hypothesis associates CP El Niño occurrence with global warming. Yeh et al. [28] demonstrated that human-induced warming weakens the Walker circulation and trade winds, leading to a shoaling and flattening of the equatorial Pacific thermocline, enhancing positive feedbacks in the central Pacific and promoting localized warming anomalies. The third hypothesis links CP El Niño occurrence to the mean-state SST patterns. Xiang et al. [29] identified a La Niña-like SST pattern emerging in the tropical Pacific around 2000, which suppressed local convective activity and shifted westerly anomalies westward, effectively inhibiting the eastward propagation of SST anomalies. The fourth hypothesis emphasizes subtropical precursors in CP El Niño occurrence. Yu et al. [30] observed that, before the occurrence of a CP El Niño event, warm SST anomalies initially form in the extratropical northeastern Pacific through the seasonal footprinting mechanism [31–33]. These warm SST anomalies then extend equatorward to the central equatorial Pacific to drive the CP El Niño event.

Changes in the frequency of the CP and EP El Niño events are also a focus [34,35]. Recent research suggests that both CP and EP El Niño events will occur much more frequently under the projected global warming scenarios in the future [28,36,37]. Specifically, weakened trade winds under global warming would flatten the equatorial Pacific thermocline while shoaling the central Pacific thermocline at the beginning of CP El Niño [28]. However, this hypothesis struggles to account for observed trends over the past 30 years [1]. In addition, since the late 20th century, the CP El Niño events have become more frequent and intense, while EP El Niño events have declined and become stronger [14,38]. It is worth mentioning that, though the global mean surface temperature plateaued since the 1990s for about a decade [39], the CP El Niño events became more frequent and intense [14,28,29,40,41]. In particular, Lee and McPhaden [40] observed a doubling in the intensity of CP El Niño events from 1982 to 2010 based on observational data, arguing that

Central Pacific thermocline shoaling could not explain the amplified strength of the CP El Niño events.

As mentioned above, many studies have noted the recent increase in the frequency of CP El Niño events [14,28,36,38]; however, the underlying mechanism is still unclear. It motivates us to focus on the increased frequency of the CP El Niño events in recent decades and figure out the mechanism. Based on the Pettitt test, we confirm a significant increase in the frequency of CP El Niño events since 2000. Between 1960 and 1999, eight CP El Niño events were recorded over a 40-year period, whereas six events were observed during the shorter 23-year span from 2000 to 2022. To explore the drivers of this increased frequency, we conduct a heat budget analysis of the upper 70 m layer in the central equatorial Pacific using ECCO2 model outputs. Our results highlight that more frequent anomalous warm zonal advection over the central equatorial Pacific contributes to more frequent warm SST (mixed layer temperature) anomalies, leading to the observed increased frequency of the CP El Niño events since 2000. This paper is organized as follows: Section 2 offers the data and methods; Section 3 confirms the significantly increased frequency of CP El Niño events since 2000; Section 4 investigates the mechanisms driving the increases in the frequency of CP El Niño events. This study culminates in Sections 5 and 6, synthesizing key conclusions and providing an interpretive analysis of the results.

2. Data and Methods

2.1. Data

El Niño events are identified according to the Ocean Niño Index (ONI) [42]. The ONI is downloaded from https://origin.cpc.ncep.noaa.gov/products/analysis_monitoring/ensostuff/ONI_v5.php (accessed on 15 June 2024).

Merged EP and CP indices are used to identify the EP and CP El Niño events. The Merged EP and CP indices are created using four widely used ENSO indices. The four ENSO indices are as follows: (1) the CP and EP indices by Yu and Kao [16], which were downloaded from https://www.ess.uci.edu/~yu/2OSC/ (accessed on 11 May 2024); (2) the C and E indices by Takahashi et al. [43]; (3) the Niño4 and Niño3 indices proposed by Yeh et al. [28]; and (4) the warm pool and cold tongue Niño indices (N_{WP} and N_{CT}) by Ren and Jin [44]. Here, we calculated the C and E indices, the Niño4 and Niño3 indices, and the N_{WP} and N_{CT} indices using ERSST V5 data [45] according to the corresponding methods. The ERSST V5 data were exclusively used to compute these indices and downloaded from https://www.psl.noaa.gov/data/gridded/data.noaa.ersst.v5.html (accessed on 4 June 2024).

The Hadley Centre Sea Ice and Sea Surface Temperature (HadISST) dataset is used to analyze the tropical Pacific SST. The HadISST is a global monthly SST and sea ice analysis developed by the Met Office Hadley Centre [46]. It has a spatial resolution of $1^\circ \times 1^\circ$ and spans from 1870 to the present. The data were downloaded from https://www.metoffice.gov.uk/hadobs/hadisst/index.html (accessed on 10 November 2024).

The Estimating the Circulation and Climate of the Ocean, Phase II (ECCO2) Cube92 model output is employed to perform a heat budget analysis in the central Equatorial Pacific. The ECCO2 is a model simulation performed based on an eddy-permitting configuration of the Massachusetts Institute of Technology general circulation model [47]. The ECCO2 estimates fit the observations using a Green's Function approach. The simulation is dynamically and kinematically consistent, with a mean horizontal resolution of 18 km and 50 depth levels to 5906 m. The ECCO2 output provides the required variables that are necessary for the heat budget analysis, including the vertical current velocity. The interpolated ECCO2 output, averaged every 3 days on a $0.25^\circ \times 0.25^\circ$ horizontal grid between 1993 and

2022, was downloaded from https://apdrc.soest.hawaii.edu/datadoc/ecco2_cube92.php (accessed on 11 May 2024).

Surface 10 m wind from the NCEP/NCAR Reanalysis I [48] is also used. The reanalysis product provides global monthly surface wind on a 2.5° × 2.5° horizontal grid since 1960. The wind data were downloaded from https://psl.noaa.gov/data/gridded/data.ncep.reanalysis.html (accessed on 17 January 2024).

2.2. Methods

2.2.1. Identification of CP and EP El Niño Events

El Niño events are identified according to the ONI criterion proposed by the National Oceanic and Atmospheric Administration (NOAA). An event is classified as the El Niño event when the ONI reaches or surpasses +0.5 °C for at least five consecutive overlapping three-month periods.

We applied the Merged EP/CP-index method, proposed by Jia and Guo (2025) [19], to determine all the EP and CP El Niño events from the identified El Niño events. This approach first standardizes the CP index, the C index, the Niño4 index, and the N_{WP} index, respectively. Then, the Merged CP index is derived as the arithmetic mean of these four standardized indices. The same procedure was used to obtain the Merged index (the time series of Merged CP and EP indices are shown in Figure S1a,b). The spatial patterns of SST anomalies regressed on the Merged EP and CP indices (Figure S1c,d) are consistent with previous studies [15,49].

Based on the Merged EP/CP-index, an El Niño event is classified as CP (EP) if the average Merged CP index value for December, January, and February exceeds (below) that of the Merged EP index.

2.2.2. Significance Tests

Multiple significance tests are used in this study. For correlation analyses, a two-tailed Student's *t*-test was applied. For spectral analyses, a red noise test is adopted. The Pettitt test is employed to identify the change point in the occurrence of CP El Niño events. The Pettitt test is a well-established method for detecting abrupt changes in time series, providing a solid theoretical foundation for variability analysis [50]. Previous studies have demonstrated that this method can effectively identify significant shifts in the mean of a time series, even when the exact timing of the change is unknown [51]. The test is based on the Mann–Whitney statistic U_t. In addition, effective degrees of freedom are used for long time series following the approach proposed by Pyper and Peterman [52].

2.2.3. The Mixed Layer Heat Budget Equation

For the mixed layer in a specific region, such as in the central equatorial Pacific, the mean mixed layer temperature T_m is controlled by

$$\frac{\partial T_m}{\partial t} = \frac{Q_0 - q_d}{\rho \cdot C_p \cdot h} - u\frac{\partial T_m}{\partial x} - v\frac{\partial T_m}{\partial y} - w_{ent}\frac{(T_m - T_d)}{h} + Res \quad (1)$$

where t is the time, Q_0 is the net surface heat flux, q_d is the downward radiative heat flux across the base of the upper layer, ρ is the seawater density, C_p is the specific heat of seawater, h is the mixed layer depth, u and v are the zonal and meridional current velocity, respectively, w_{ent} is the entrainment rate of cold seawater from below the mixed layer, T_d is the seawater temperature below the mixed layer, and Res is the residual containing the contribution from mixing [53,54]. $\frac{\partial T_m}{\partial t}$ on the left hand of Equation (1) is the temperature tendency term, while $\frac{Q_0 - q_d}{\rho \cdot C_p \cdot h}$, $-u\frac{\partial T_m}{\partial x}$, $-v\frac{\partial T_m}{\partial y}$, $-w_{ent}\frac{(T_m - T_d)}{h}$, and Res on the right hand are the surface thermal forcing term, the zonal temperature advection term, the meridional temperature advection term, the vertical entrainment term, and the residual term, respectively.

In this study, we use a fixed mixed layer base instead of a varying one. The study region (5° S–5° N, 170° W–122° W) is located in the central equatorial Pacific, where the seasonality of mixed layer depth is much weaker than in the midlatitudes. The mixed layer in the study region is around 50–100 m [55,56]. Here, we chose the upper 70 m layer as the mixed layer. The mean temperature of the upper 70 m layer in the study region is consistent with the regional mean SST, as shown in Figure S2. Considering the 70 m layer is much thicker, we take q_d as 0. In other words, we think that the upper 70 m layer absorbs the whole surface net heat flux. Using a fixed mixed layer, the w_{ent} follows

$$\begin{cases} w_{ent} = w_d,\ w_d > 0 \\ w_{ent} = 0,\ w_d \leq 0 \end{cases} \tag{2}$$

where w_d is the vertical current velocity at the base of the fixed mixed layer, and upward positive. The heat budget analysis is performed based on the ECCO2 model output.

3. Increase in CP El Niño Frequency Since 2000

During the study period between 1960 and 2022, 21 El Niño events are identified according to the ONI criterion (see Section 2.2 Methods). Furthermore, based on the Merged EP/CP-index method, among the identified El Niño events, 14 were classified as CP El Niño, while the remaining 7 were categorized as EP El Niño. The results are listed in Table 1. More CP El Niño events are observed compared with EP El Niño events.

Table 1. All the CP and EP El Niño events between 1960 and 2022, as identified by the Merged EP/CP-index method (see Section 2.2 Methods).

	CP El Niño Events	EP El Niño Events
Years	1963–1964; 1965–1966; 1968–1969; 1969–1970; 1977–1978; 1979–1980; 1987–1988; 1994–1995; 2002–2003; 2004–2005; 2006–2007; 2009–2010; 2014–2015; 2018–2019	1972–1973; 1976–1977; 1982–1983; 1986–1987; 1991–1992; 1997–1998; 2015–2016

The frequency of CP El Niño events increased since 2000. Figure 1a displays the frequency of CP El Niño events from 1960 to 2022. If a CP El Niño event occurs, the occurrence is denoted as 1; otherwise, the occurrence is 0. A notable increase in CP El Niño frequency seems to emerge since 2000. During 1960–1999, eight CP El Niño events are recorded over the 40 years. Meanwhile, during the shorter 23-year span of 2000–2022, six CP El Niño events are observed. To investigate the change in the frequency, we applied the Pettitt test to detect abrupt changes in the CP El Niño events, as shown in Figure 1b. The CP El Niño events exhibited significant abrupt changes around 2000. Specifically, the CP El Niño events underwent a significant abrupt change in 2001, shifting from a declining trend pre-2001 to a rising trend thereafter.

CP El Niño events experienced significant changes since 2000. To further investigate the frequency of CP El Niño events, Figures S3–S6 present the number of events within 7-year, 9-year, 11-year, and 13-year moving windows, respectively. The results are consistent with the significant abrupt changes observed after 2000. Specifically, CP El Niño events exhibited significant abrupt changes across various moving window lengths. Within a 7-year moving window, a notable shift occurred in 2000 ($p < 0.05$), changing from a declining trend prior to 2000 to an increasing trend thereafter. In the 9-year moving window, a significant change was detected between 1999 and 2001 ($p < 0.05$). For the 11-year window, the change occurred in 1998 ($p < 0.05$), while in the 13-year window, a significant abrupt shift was observed in 1999 ($p < 0.05$). Hence, using 2000 as the change point, we will analyze the changes in CP El Niño events during 1960–1999 and 2000–2022. A statistical analysis of the number of CP El Niño events for each decade from the 1960s to the 2010s further confirms the abrupt changes in the CP El Niño frequency after 2000 (see Figure S7). The result is consistent with previous studies [14,38].

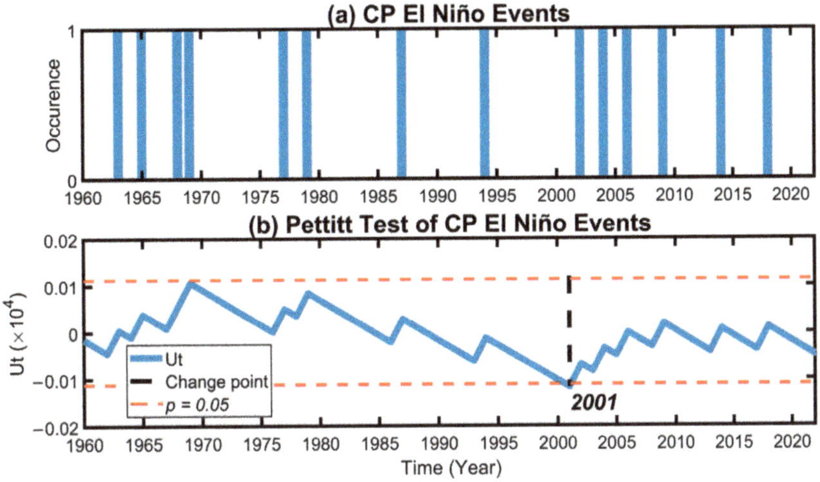

Figure 1. Statistical analysis and Pettitt test of the frequency of CP El Niño events. (**a**) The number of CP El Niño events; (**b**) Pettitt test for the number of CP El Niño events, where the solid line denotes the test statistic, the black dashed line indicates the detected change point, and the red dashed line represents the 95% confidence level.

Since 2000, the period of CP El Niño's interannual variability shortened. We first examine the differences in the interannual variability of the CP El Niño before and since 2000. Figure 2a presents the time series of the Merged CP index. The power spectrum results of the Merged CP index during 1960–1999 and 2000–2022 are shown in Figure 2b,c, respectively. The Merged CP index during 1960–1999 exhibited the most significant period at 2.5, 3.6, and 4.4 years (passing the 99% confidence level, Figure 2b). Since 2000, the amplitude of the interannual variability in CP El Niño did not show obvious changes. However, the most significant period of the interannual variability of the Merged CP index shortened and shifted to 2–3 years since 2000, as shown in Figure 2c. The period of the CP El Niño events shortens from 4–5 years to 2–3 years. In summary, the interannual variability in the CP El Niño exhibited a shortened period. Consistent results were also obtained from the wavelet power spectrum of the Merged CP index, as illustrated in Figure S8.

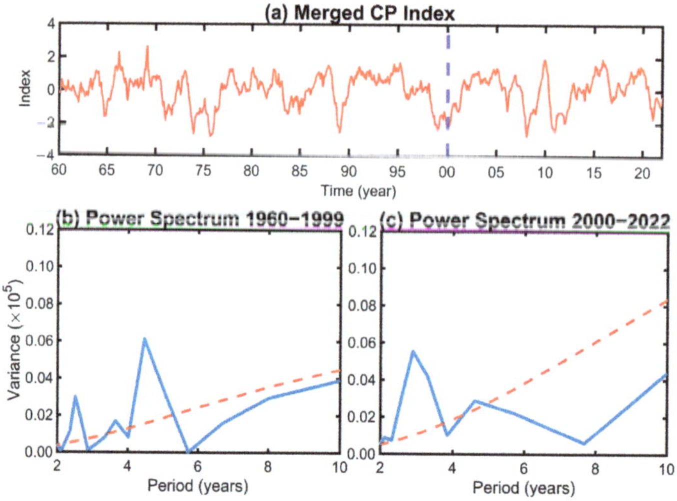

Figure 2. The time series and power spectrum analysis of the Merged CP index. (**a**) The time series of the Merged CP index from 1960 to 2022. The blue dashed line represents the year 2000; (**b**) Power spectrum analysis of the Merged CP index in 1960–1999. The blue solid line represents the power spectrum, and the red dashed line offers a 99% significance level; (**c**) Same as (**b**), but during 2000–2022.

4. Mechanisms

4.1. More Frequent Warm SST Anomalies in the Key Region

To investigate the increased frequency of CP El Niño events since 2000, we first performed a composite analysis of SST anomalies across three evolutionary phases: the onset, developing, and mature phases. Following previous studies [57–60], February–April (FMA) and April–June (AMJ) are chosen to indicate the onset phase, June-August (JJA) and August-October (ASO) are taken to represent the developing phase, and October-December (OND) and December-February (D(0)JF(+1)) are picked to reflect the mature phase. The composite SST anomalies from FMA(0) to D(0)JF(+1) during the CP El Niño events between 1960 and 2022 are illustrated in Figure 3a–f, respectively. Significant warm SST anomalies first appear in the central equatorial Pacific in JJA(0) of the developing phase (Figure 3c). The specific region with significant SST anomalies is 5° S–5° N, 170° W–122° W (the black box in Figure 3c). Then, the significant warm SST anomalies amplify and extend eastward, as from Figure 3d–f. The specific region of 5° S–5° N, 170° W–122° W is essential for the developing phase of the CP El Niño. We take it as the key region for the CP El Niño.

The SST anomalies in the key region are highly linked to CP El Niño events during the developing phase. Figure 4a,b present the time series of regional mean SST anomalies in the key region and the Merged CP index during the developing phase, respectively. The two time series seem to have the same phase change. A lead–lag correlation analysis further confirms the close relationship. As shown in Figure 4c, the largest positive correlation coefficient reaches 0.75 at zero leads or lags, passing the 99% confidence level. It suggests that the SST anomalies in the key region and the Merged CP index during the developing phase of the CP El Niño events exhibit a significant simultaneous correlation.

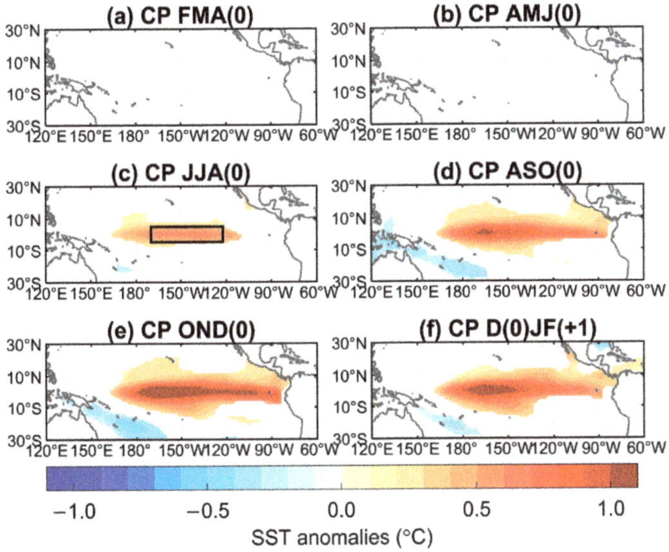

Figure 3. Composite of SST anomalies during the onset, developing, and mature phases of the CP El Niño event (1960–2022). (**a**,**b**) Composite of SST anomalies during the onset phase (February to June) of the CP El Niño event. The black box represents the key region, which extends from 5° S to 5° N, 170° W to 122° W; (**c**,**d**) Same as (**a**,**b**), but for the developing phase (June to October); (**e**,**f**) Same as (**a**,**b**), but for the mature phase (October to February of the following year). Color shading highlights significant SST anomalies that exceed the 95% confidence level.

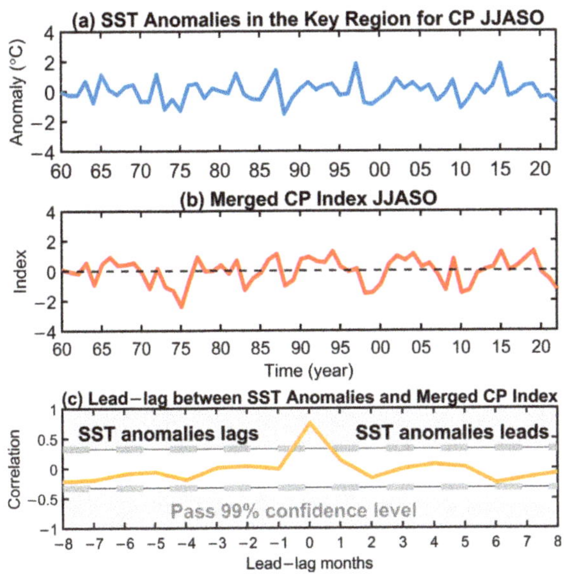

Figure 4. Regional mean SST anomalies in the key region for CP El Niño and the Merged CP index during the developing phase. (**a**) The time series of regional mean SST anomalies in the key region for CP El Niño during the developing phase. (**b**) The time series of the Merged CP index during the developing phase. The black dashed line indicates the zero line (y = 0). (**c**) Lead–lag correlation analysis between the time series of regional mean SST anomalies in the key region for CP El Niño and the Merged CP index during the developing phase. The grey dashed lines represent the 99% confidence levels.

Notably, warm SST anomalies in the key region are more frequently observed during 2000–2022 compared to 1960–1999. Figure 5a shows the time series of regional mean SST anomalies in the key region from 1960 to 2022 during the developing phase. Between 1960 and 1999, there were 14 years that presented large and persistent warm SST anomalies in the key region over 40 years. In contrast, between 2000 and 2022, there were 9 years in which large and persistent warm SST anomalies emerged in just 23 years. The increased frequency of the warm SST anomalies in the key region since 2000 is further confirmed by a wavelet analysis. Figure 5b shows the wavelet power spectrum of the regional mean SST anomalies. The regional mean SST anomalies in the key region exhibit two dominant periods before and after 2000. During 1960–1999, the SST anomalies showed a significant period of 2–5 years. However, during 2000–2022, the significant period shifts to 2–3 years. The more frequent warm SST anomalies in the key region during the developing phase of the CP El Niño are consistent with the shortened period of the interannual variability of the Merged CP index in Figure 2c since 2000. The more frequent warm SST anomalies in the key region drive the observed increased frequency of the CP El Niño events since 2000.

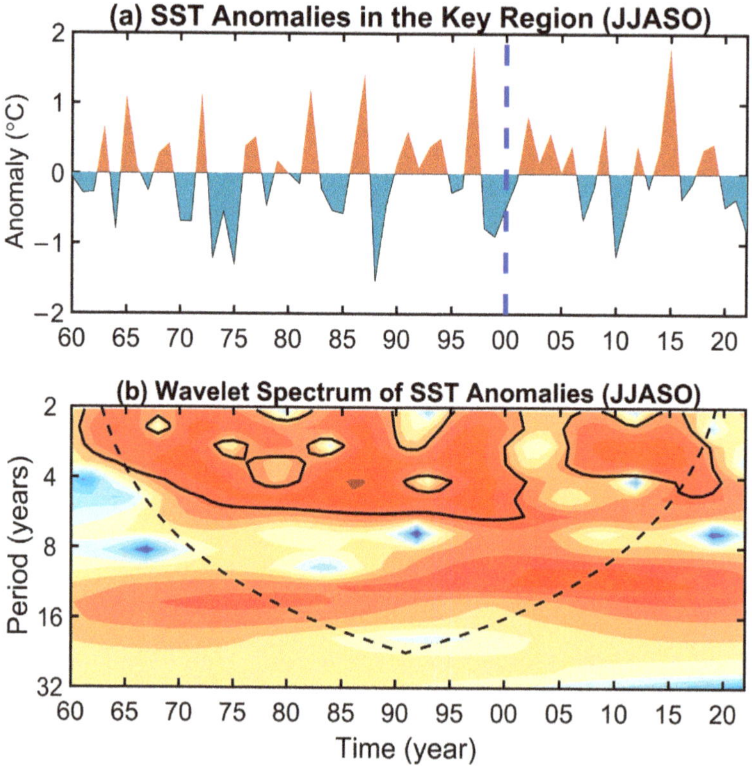

Figure 5. Regional mean SST anomalies and wavelet power spectrum in the key region for CP El Niño during the developing phase. (**a**) The time series of regional mean SST anomalies in the key region for CP El Niño during the developing phase. Red shading denotes warm SST anomalies, blue shading denotes cold SST anomalies, and the blue dashed line marks the year 2000; (**b**) Wavelet power spectrum of regional mean SST anomalies in the key region for CP El Niño during the developing phase (color shading). The solid contour lines denote the 95% significance level, The dashed contour line indicates the cone of influence.

4.2. Role of Anomalous Warm Zonal Advection

So, why are warm SST anomalies in the key region more frequently observed during the developing phase since 2000? To find out the answer, we conducted a heat budget analysis of the mixed layer in the key region based on the ECCO2 model output. The ECCO2 output spans from 1992 to 2022. Here, we used a fixed mixed layer, and the mixed layer depth is chosen to be 70 m according to previous studies [61]. The two datasets, HadISST and ECCO2, also exhibit nearly identical time series of monthly regional mean SST anomalies in the key region, as shown in the yellow line and the blue line in Figure 6, respectively. The time series of the regional mean SST in the key region from HadISST and the 70 m depth-averaged regional mean temperature in the key region from ECCO2 are offered in the yellow line and the red line in Figure 6, respectively. Clearly, the two time series show nearly the same phase change. This suggests that the SST and the subsurface temperature in the Central Pacific of the ECCO2 output are consistent with the SST of the HadISST. This guarantees that performing the heat budget analysis of the upper 70 m layer in the Central Pacific based on the ECCO2 output is valid.

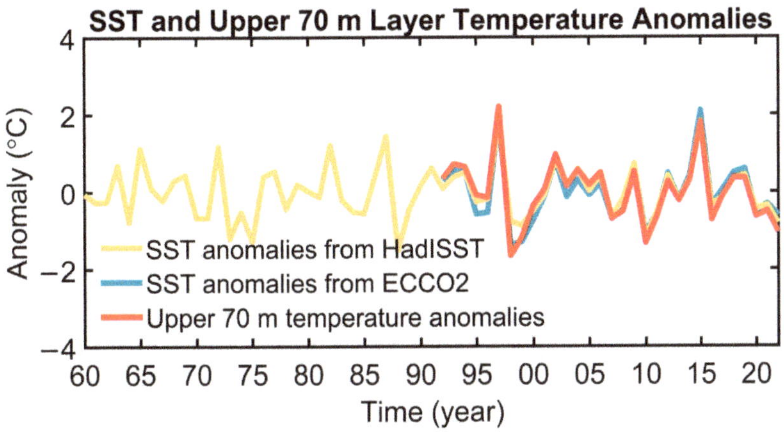

Figure 6. Time series of regional mean SST anomalies in the key region during the developing phase, based on HadISST (yellow) and ECCO2 (blue) datasets, along with the upper 70 m layer regional mean temperature from ECCO2 (red).

The results of the heat budget analysis of the upper 70 m layer in the key region during the developing phase (JJASO) are illustrated in Figure 7. The temperature tendency (Figure 7a) mainly follows the zonal temperature advection (Figure 7c). Their correlation coefficient is 0.96, passing the 95% confidence level. The correlation coefficient between the temperature tendency and surface thermal forcing is −0.67. In comparison, the correlation with meridional temperature advection is 0.81, the vertical entrainment is 0.81, and the residual term is −0.80. Other terms also show significant correlation coefficients, but they are less important than zonal temperature advection. Surface thermal forcing exhibits no significant changes over time. Zonal temperature advection is the dominant process during the developing phase of CP El Niño events. In contrast, meridional temperature advection plays a minor role in the warming during this phase and shows a further decline after 2000. Vertical entrainment remains relatively steady throughout the period. The residual term associated with heat flux represents a combination of mixing. In all, on the interannual time scale, the SST, as well as mixed layer temperature variability in the key region is determined by zonal temperature advection.

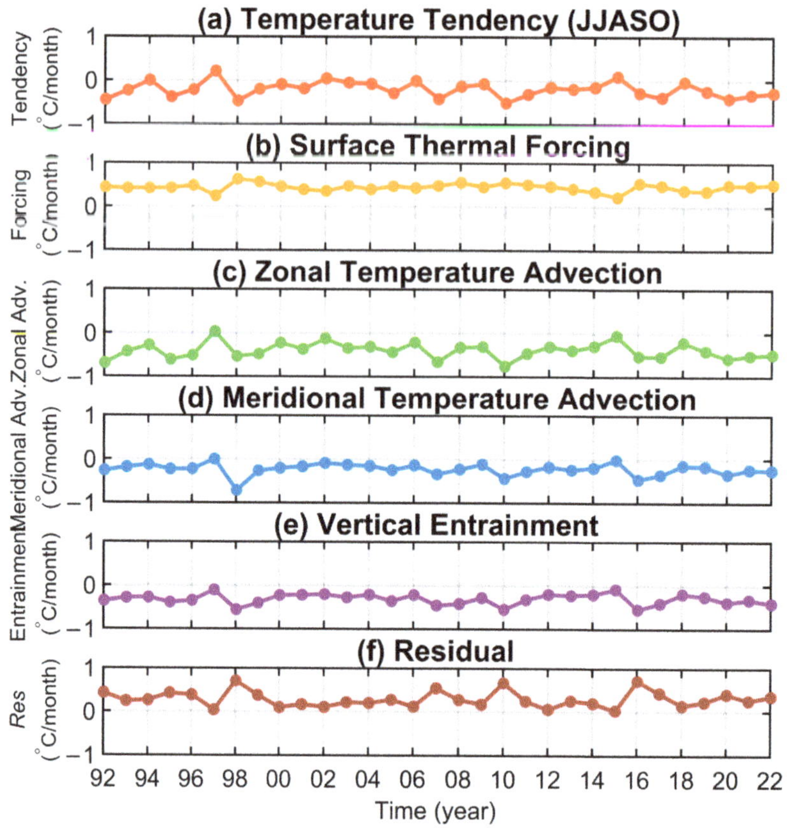

Figure 7. The heat budget analysis of the upper 70 m layer in the key region based on the ECCO2 model output between 1992 and 2022. (**a**) The monthly depth-averaged regional mean temperature tendency term. (**b**) The surface thermal forcing term, (**c**) the zonal temperature advection term, (**d**) the meridional temperature advection term, (**e**) the vertical entrainment term, and (**f**) the residual term.

The more frequent warm SST anomalies in the key region during the developing phase are induced by the more frequent anomalous warm zonal advection. Figure 8a shows the time series of regional mean zonal advection anomalies in the key region from 1992 to 2022 during the developing phase. Figure 8b,c present the time series of regional mean SST anomalies in the key region and the Merged CP index during the developing phase, respectively. The warm zonal advection and SST anomalies appear to exhibit in-phase changes. In the key region, correlation analysis reveals that the correlation coefficient was 0.80 during 2000–2022, passing the 99% confidence level. Between 2000 and 2022, there were 9 years in large and persistent warm zonal advection anomalies which emerged in just 23 years. Specifically, anomalous warm zonal advection is observed in the years 2002, 2004, 2006, 2008, 2009, 2012, 2014, 2015, and 2018, consistent with the results shown in Figure 5a.

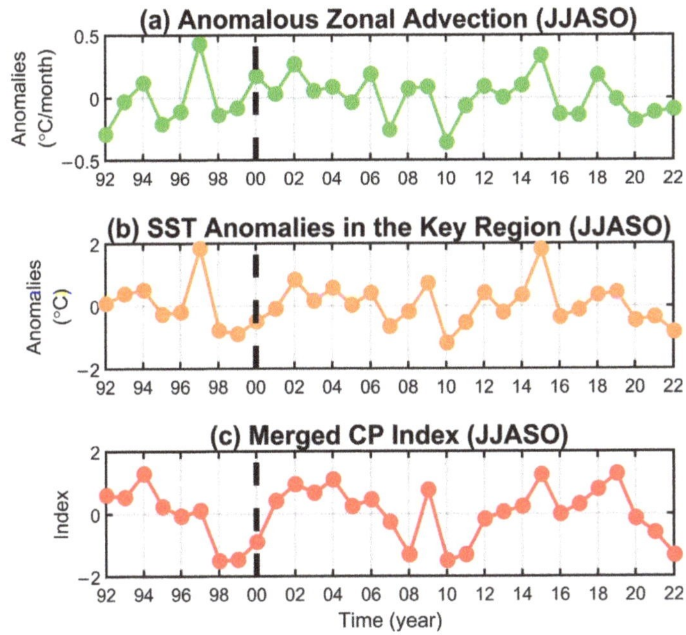

Figure 8. Time series of regional mean zonal advection and SST anomalies in the key region and Merged CP index during the developing phase. (**a**) The zonal temperature advection term from the heat budget analysis of the upper 70 m layer in the key region, based on ECCO2 model output from 1992 to 2022; (**b**) The time series of regional mean SST anomalies in the key region for CP El Niño during the developing phase; (**c**) The time series of the Merged CP index during the developing phase. The black dashed line marks the year 2000.

5. Conclusions

This study investigates the recent increase in the frequency of CP El Niño events. Based on the Merged EP/CP-Index method, we identify all the CP El Niño events between 1960 and 2022. There are 14 CP El Niño events over the 63 years. A Pettitt test is used to analyze the change in the frequency of CP El Niño events during 1960–2022. The results confirm a significant increase in the frequency of CP El Niño events since 2000. Over the 40 years between 1960 and 1999, eight CP El Niño events appeared. In contrast, over the 23 years between 2000 and 2022, six CP El Niño events are seen. In addition, the period of the CP El Niño events shortens from 4–5 years to 2–3 years.

A composite analysis of the SST anomalies during the CP El Niño event reveals that the key region (5° S–5° N, 170° W–122° W) in the central equatorial Pacific is essential for the development of CP El Niño events. The increased frequency of CP El Niño events is closely related to more frequent warm SST anomalies in this key region during the developing phase of CP El Niño events (i.e., June to October). The two datasets, HadISST and ECCO2, also exhibit nearly identical time series of monthly regional mean SST anomalies in the key region. Furthermore, a heat budget analysis of the upper 70 m layer in the key region based on the ECCO2 output suggests that the SST variability in the central equatorial Pacific during the developing phase is dominated by zonal temperature advection. The more frequent warm SST anomalies are caused by frequent anomalous warm zonal advection. In all, we propose that the frequent anomalous warm zonal advection since 2000 causes the more frequent warm SST anomalies, and finally drives the observed increase in the CP El Niño event frequency since 2000.

6. Discussion

The frequency of the EP El Niño events has declined since the 2000s. Figure 9a,b display the number of EP El Niño events. A notable decline in EP El Niño frequency is evident since 2000, with only one EP El Niño event occurring over the subsequent two decades (Figure 9a). To determine the timing of the frequency changes, we applied the Pettitt test to detect abrupt changes in the number of EP El Niño events, as shown in Figure 9b. EP El Niño events also exhibited significant abrupt changes since 2000, yet their frequency trends diverged markedly from CP El Niño events. Specifically, the EP El Niño events showed a statistically significant abrupt change in 1997 ($p < 0.05$), transitioning from a significant increasing trend pre-1997 to a declining trend since 1997. These results reveal opposing frequency characteristics between EP and CP El Niño events.

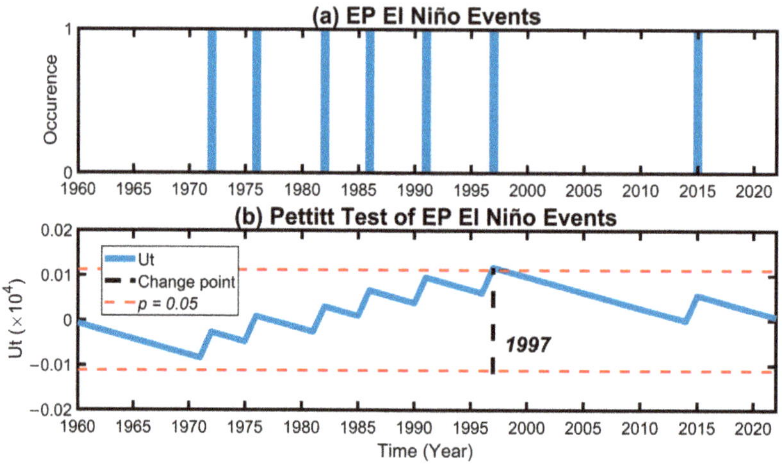

Figure 9. Statistical analysis and Pettitt test of the frequency of EP El Niño events. (**a**) The number of EP El Niño events; (**b**) Pettitt test for the number of EP El Niño events, where the solid line denotes the test statistic, the black dashed line indicates the detected change point, and the red dashed line represents the 95% confidence level.

Consistent with Shin et al. (2022) [14], our results also indicate a marked increase in CP El Niño frequency since 2000. Previous studies also reported that there has been an increase in extreme El Niño events in recent years [34]. Although our study examines the trends and changes in EP and CP El Niño events, we do not specifically analyze their intensities. We classify the two types of El Niño into four intensity levels: super, strong, moderate, and weak, based on the ONI, following L'Heureux et al. [62]. Table S1 shows the number of EP and CP El Niño events at each intensity. There are three, one, two, and one EP El Niño events classified as super, strong, moderate, and weak, respectively. Meanwhile, the number of CP El Niño events in these categories is zero, two, five, and seven, respectively. Figure S9 shows the proportion of events at each intensity. From the distribution of different intensities of EP and CP El Niño events, we find that EP El Niño events are predominantly super El Niño events, accounting for 43% (Figure S9a). In contrast, 50% of CP El Niño events are weak, followed by 36% classified as moderate, with no super CP El Niño events observed (Figure S9b). As the current number of EP and CP El Niño events of different intensities and types is too limited to analyze trends and variability, further research is essential to incorporate longer-term data or model experiments, which would help investigate the characteristics of El Niño frequency under different intensities. Moreover, a complementary relationship between the decadal frequencies of CP El Niño events is evident in Figures S7 and S10: when the frequency of EP El Niño events is at its lowest, CP El Niño events reach a peak, while the total number of events appears to remain relatively constant. Therefore, further investigation is warranted to determine whether these two types of El Niño events represent different manifestations of the same underlying decadal variability. The zonal temperature advection during the developing phase of CP El Niño events might be linked to anomalous zonal winds over the key region.

Supplementary Materials: The following supporting information can be downloaded at: https://www.mdpi.com/article/10.3390/atmos16060654/s1, Figure S1: The Merged EP/CP index and the corresponding spatial patterns of winter Sea Surface Temperature (SST) anomalies for EP and CP El Niño events are presented. (a) and (b) show the time series of the monthly merged EP and CP indices from 1960 to 2022, respectively. (c) and (d) display the composite spatial patterns of SST anomalies (color shading, in °C) for EP and CP El Niño events in December, January, and February (0/1). Among the 21 El Niño events, 7 EP and 14 CP El Niño events are used in the composites (see Table 1). This analysis follows the methodology of Jia and Guo (2025); Figure S2: Time series of regional mean SST anomalies (blue) in the key region during the developing phase and the upper 70 m layer regional mean temperature (red); Figure S3: Statistical analysis and Pettitt test of the frequency of CP El Niño events. (a) The number of CP El Niño events within a 7-year moving window; (b) Pettitt test for the number of CP El Niño events within a 7-year moving window, where the solid line denotes the test statistic, the black dashed line indicates the detected change point, and the magenta, blue, and yellow dashed lines represent the 99%, 95%, and 90% confidence levels, respectively; Figure S4: Statistical analysis and Pettitt test of the frequency of CP El Niño events. (a) The number of CP El Niño events within a 9-year moving window; (b) Pettitt test for the number of CP El Niño events within a 9-year moving window, where the solid line denotes the test statistic, the black dashed line indicates the detected change point, and the red dashed line represents the 95% confidence level; Figure S5: Statistical analysis and Pettitt test of the frequency of CP El Niño events. (a) The number of CP El Niño events within a 11-year moving window; (b) Pettitt test for the number of CP El Niño events within a 11-year moving window, where the solid line denotes the test statistic, the black dashed line indicates the detected change point, and the magenta, blue, and yellow dashed lines represent the 99%, 95%, and 90% confidence levels, respectively; Figure S6: Statistical analysis and Pettitt test of the frequency of CP El Niño events. (a) The number of CP El Niño events within a 13-year moving window; (b) Pettitt test for the number of CP El Niño events within a 13-year moving window, where the solid line denotes the test statistic, the black dashed line indicates the detected change point, and the magenta, blue, and yellow dashed lines represent the 99%, 95%, and 90% confidence levels, respectively; Figure S7: The number of occurrences of the CP El Niño events in each decade from the 1960s to the 2010s; Figure S8: The real part of wavelet coefficients for the Merged CP index, with black soild lines indicating the 95% significance level; Table S1: All the El Niño events with four different intensities between 1960 and 2022 according to the ONI-based criterion and their types decided by the Merged EP/CP-index methods; Figure S9: Proportion of two types of El Niño events with different intensities. (a) Proportions of super El Niño, strong El Niño, moderate El Niño, and weak El Niño events within EP El Niño events; (b) Same as (a), but for CP El Niño events; Figure S10: The number of occurrences of the EP El Niño events in each decade from the 1960s to the 2010s.

Author Contributions: Conceptualization, L.J. and Y.G.; methodology, L.J.; software, L.J.; validation, L.J. and Y.G.; formal analysis, L.J.; investigation, L.J.; resources, L.J.; data curation, L.J.; writing—original draft preparation, L.J.; writing—review and editing, L.J. and Y.G.; visualization, L.J. and Y.G.; supervision, Y.G.; project administration, Y.G.; funding acquisition, Y.G. All authors have read and agreed to the published version of the manuscript.

Funding: This research was funded by the National Natural Science Foundation of China, grant number: 42106019. Bureau of Science and Technology of Zhoushan, the Zhoushan Science and Technology Project—Zhejiang Ocean University Special Project, grant number: 2022C41020. The Scientific Research Initiation Fund of the Zhejiang Ocean University, grant number: GK 11034150220003.

Institutional Review Board Statement: Not applicable.

Informed Consent Statement: Not applicable.

Data Availability Statement: The ONI is downloaded from https://origin.cpc.ncep.noaa.gov/products/analysis_monitoring/ensostuff/ONI_v5.php (accessed on 15 June 2024). The EP and CP indices were downloaded from https://www.ess.uci.edu/~yu/2OSC/ (accessed on 11 May 2024). The ERSST V5 data from https://www.psl.noaa.gov/data/gridded/data.noaa.ersst.v5.html (accessed on 4 June 2024). The HadISST data were downloaded from https://www.metoffice.gov.uk/hadobs/hadisst/index.html (accessed on 10 November 2024). The NCEP/NCAR Reanalysis I dataset is available for download at: https://psl.noaa.gov/data/gridded/data.ncep.reanalysis.html (accessed on 17 January 2024). The interpolated ECCO2 output was downloaded from https://apdrc.soest.hawaii.edu/datadoc/ecco2_cube92.php (accessed on 11 May 2024).

Acknowledgments: Jia and Guo would like to thank Ji Qi for his valuable help and suggestions in this study. Qi helped perform the heat budget analysis with the ECCO2 output. Special thanks are extended to Editors for their valuable guidance and support throughout the revision and review process. We are also grateful to the three anonymous reviewers for their insightful comments, thoughtful questions, and constructive suggestions, which significantly improved our work. We deeply appreciate their time and effort in reviewing our manuscript.

Conflicts of Interest: The authors declare no conflicts of interest.

Abbreviations

The following abbreviations are used in this manuscript:

ENSO	El Niño–Southern Oscillation
SST	Sea surface temperature
EP	Eastern Pacific
CP	Central Pacific
ONI	Ocean Niño Index
N_{WP}	Warm pool Niño index
N_{CT}	Cold tongue Niño index
HadISST	The Hadley Centre Sea Ice and Sea Surface Temperature
FMA	February, March, and April
AMJ	April, May, and June
JJA	June, July, and August
ASO	August, September, and October
OND	October, November, and December
DJF	December, January, and February

References

1. Newman, M.; Shin, S.I.; Alexander, M.A. Natural variation in ENSO flavors. *Geophys. Res. Lett.* **2011**, *38*, L14705. [CrossRef]
2. Bjerknes, J. Atmospheric Teleconnections from the Equatorial Pacific1. *Mon. Weather Rev.* **1969**, *97*, 163–172. [CrossRef]
3. Torrence, C.; Webster, P.J. Interdecadal Changes in the ENSO–Monsoon System. *J. Clim.* **1999**, *12*, 2679–2690. [CrossRef]
4. Nicholls, N. Predictability of Interannual Variations of Australian Seasonal Tropical Cyclone Activity. *Mon. Weather Rev.* **1985**, *113*, 1144–1149. [CrossRef]
5. Rasmusson, E.M.; Carpenter, T.H. Variations in Tropical Sea Surface Temperature and Surface Wind Fields Associated with the Southern Oscillation/El Niño. *Mon. Weather Rev.* **1982**, *110*, 354–384. [CrossRef]
6. Jin, F.F.; Kug, J.S.; An, S.-I. Two Types of El Niño Events: Cold Tongue El Niño and Warm Pool El Niño. *J. Clim.* **2009**, *22*, 1499–1515. [CrossRef]
7. Troup, A. The 'southern oscillation'. *Q. J. R. Meteorol. Soc.* **1965**, *91*, 490–506. [CrossRef]
8. Kim, H.M.; Webster, P.J.; Curry, J.A. Impact of shifting patterns of Pacific Ocean warming on North Atlantic tropical cyclones. *Science* **2009**, *325*, 77–80. [CrossRef]
9. Wang, C.; Deser, C.; Yu, J.-Y.; DiNezio, P.; Clement, A. El Niño and Southern Oscillation (ENSO): A Review. In *Coral Reefs of the Eastern Tropical Pacific*; Coral Reefs of the World; Springer: Berlin/Heidelberg, Germany, 2017; pp. 85–106.

10. Alexander, M.A.; Bladé, I.; Newman, M.; Lanzante, J.R.; Lau, N.-C.; Scott, J.D. The atmospheric bridge: The influence of ENSO teleconnections on air–sea interaction over the global oceans. *J. Clim.* **2002**, *15*, 2205–2231. [CrossRef]
11. Jiang, T.; Kundzewicz, Z.W.; Su, B. Changes in monthly precipitation and flood hazard in the Yangtze River Basin, China. *Int. J. Climatol. A J. R. Meteorol. Soc.* **2008**, *28*, 1471–1481. [CrossRef]
12. Changnon, S.A. Impacts of 1997–98 EI Niño–generated weather in the United States. *Bull. Am. Meteorol. Soc.* **1999**, *80*, 1819–1828. [CrossRef]
13. Alizadeh-Choobari, O. Contrasting global teleconnection features of the eastern Pacific and central Pacific El Niño events. *Dyn. Atmos. Ocean.* **2017**, *80*, 139–154. [CrossRef]
14. Shin, N.Y.; Kug, J.-S.; Stuecker, M.F.; Jin, F.F.; Timmermann, A.; Kim, G.-I. More frequent central Pacific El Niño and stronger eastern pacific El Niño in a warmer climate. *NPJ Clim. Atmos. Sci.* **2022**, *5*, 101. [CrossRef]
15. Ashok, K.; Behera, S.K.; Rao, S.A.; Weng, H.; Yamagata, T. El Niño Modoki and its possible teleconnection. *J. Geophys. Res.* **2007**, *112*, C11007. [CrossRef]
16. Yu, J.Y.; Kao, H.Y. Contrasting Eastern-Pacific and Central-Pacific Types of ENSO. *J. Clim.* **2009**, *22*, 615–632. [CrossRef]
17. Kim, S.T.; Yu, J.-Y. Reversed Spatial Asymmetries between El Niño and La Niña and Their Linkage to Decadal ENSO Modulation in CMIP3 Models. *J. Clim.* **2011**, *24*, 5423–5434. [CrossRef]
18. Larkin, N.K.; Harrison, D. Global seasonal temperature and precipitation anomalies during El Niño autumn and winter. *Geophys. Res. Lett.* **2005**, *32*. [CrossRef]
19. Jia, L.; Guo, Y. Revisiting the differential impacts of Eastern-Pacific and Central-Pacific El Niño on late autumn to winter precipitation over South China. *Dyn. Atmos. Ocean.* **2025**, *110*, 101557. [CrossRef]
20. Yu, J.Y.; Zou, Y.h.; Kim, S.T.; Lee, T. The changing impact of El Niño on US winter temperatures. *Geophys. Res. Lett.* **2012**, *39*, L15702. [CrossRef]
21. Schopf, P.S.; Suarez, M.J. Vacillations in a Coupled Ocean–Atmosphere Model. *J. Atmos. Sci.* **1988**, *45*, 549–566. [CrossRef]
22. Jin, F.-F. An equatorial ocean recharge paradigm for ENSO. Part II: A stripped-down coupled model. *J. Atmos. Sci.* **1997**, *54*, 830–847. [CrossRef]
23. Weisberg, R.H.; Wang, C. A western Pacific oscillator paradigm for the El Niño-Southern Oscillation. *Geophys. Res. Lett.* **1997**, *24*, 779–782. [CrossRef]
24. Picaut, J.; Masia, F.; du Penhoat, Y. An advective-reflective conceptual model for the oscillatory nature of the ENSO. *Science* **1997**, *277*, 663–666. [CrossRef]
25. Battisti, D.S.; Hirst, A.C. Interannual Variability in a Tropical Atmosphere–Ocean Model: Influence of the Basic State, Ocean Geometry and Nonlinearity. *J. Atmos. Sci.* **1989**, *46*, 1687–1712. [CrossRef]
26. Jin, F.F. An equatorial ocean recharge paradigm for ENSO. Part I: Conceptual model. *J. Atmos. Sci.* **1997**, *54*, 811–829. [CrossRef]
27. Wang, C.Z.; Weisberg, R.H.; Virmani, J.I. Western Pacific interannual variability associated with the El Niño-Southern Oscillation. *J. Geophys. Res. Ocean.* **1999**, *104*, 5131–5149. [CrossRef]
28. Yeh, S.W.; Kug, J.S.; Dewitte, B.; Kwon, M.H.; Kirtman, B.P.; Jin, F.-F. El Niño in a changing climate. *Nature* **2009**, *461*, 511–514. [CrossRef]
29. Xiang, B.Q.; Wang, B.; Li, T. A new paradigm for the predominance of standing central Pacific warming after the late 1990s. *Clim. Dyn.* **2013**, *41*, 327–340. [CrossRef]
30. Yu, J.-Y.; Kao, H.-Y.; Lee, T. Subtropics-related interannual sea surface temperature variability in the central equatorial Pacific. *J. Clim.* **2010**, *23*, 2869–2884. [CrossRef]
31. Vimont, D.J.; Alexander, M.; Fontaine, A. Midlatitude excitation of tropical variability in the Pacific: The role of thermodynamic coupling and seasonality. *J. Clim.* **2009**, *22*, 518–534. [CrossRef]
32. Vimont, D.J.; Wallace, J.M.; Battisti, D.S. The seasonal footprinting mechanism in the Pacific: Implications for ENSO. *J. Clim.* **2003**, *16*, 2668–2675. [CrossRef]
33. Vimont, D.J.; Battisti, D.S.; Hirst, A.C. Footprinting: A seasonal connection between the tropics and mid-latitudes. *Geophys. Res. Lett.* **2001**, *28*, 3923–3926. [CrossRef]
34. Cai, W.; Borlace, S.; Lengaigne, M.; Van Rensch, P.; Collins, M.; Vecchi, G.; Timmermann, A.; Santoso, A.; McPhaden, M.J.; Wu, L. Increasing frequency of extreme El Niño events due to greenhouse warming. *Nat. Clim. Change* **2014**, *4*, 111–116. [CrossRef]
35. Gan, R.Y.; Liu, Q.; Huang, G.; Hu, K.M.; Li, X.C. Greenhouse warming and internal variability increase extreme and central Pacific El Niño frequency since 1980. *Nat. Commun.* **2023**, *14*, 394. [CrossRef] [PubMed]
36. Chen, M.Y.; Yu, J.Y.; Wang, X.; Chen, S. Distinct onset mechanisms of two subtypes of CP El Niño and their changes in future warming. *Geophys. Res. Lett.* **2021**, *48*, e2021GL093707. [CrossRef]
37. Ng, B.; Cai, W.; Cowan, T.; Bi, D. Impacts of low-frequency internal climate variability and greenhouse warming on El Niño–Southern Oscillation. *J. Clim.* **2021**, *34*, 2205–2218. [CrossRef]
38. Freund, M.B.; Henley, B.J.; Karoly, D.J.; McGregor, H.V.; Abram, N.J.; Dommenget, D. Higher frequency of Central Pacific El Niño events in recent decades relative to past centuries. *Nat. Geosci.* **2019**, *12*, 450–455. [CrossRef]

39. Solomon, S.; Rosenlof, K.H.; Portmann, R.W.; Daniel, J.S.; Davis, S.M.; Sanford, T.J.; Plattner, G.-K. Contributions of stratospheric water vapor to decadal changes in the rate of global warming. *Science* **2010**, *327*, 1219–1223. [CrossRef]
40. Lee, T.; McPhaden, M.J. Increasing intensity of El Niño in the central-equatorial Pacific. *Geophys. Res. Lett.* **2010**, *37*, L14603. [CrossRef]
41. Chung, P.H.; Li, T. Interdecadal relationship between the mean state and El Niño types. *J. Clim.* **2013**, *26*, 361–379. [CrossRef]
42. Barnston, A.G.; Chelliah, M.; Goldenberg, S.B. Documentation of a highly ENSO-related SST region in the equatorial Pacific. *Atmos.-Ocean (Can. Meteorol. Oceanogr. Soc.)* **1997**, *35*, 367. [CrossRef]
43. Takahashi, K.; Montecinos, A.; Goubanova, K.; Dewitte, B. ENSO regimes: Reinterpreting the canonical and Modoki El Niño. *Geophys. Res. Lett.* **2011**, *38*, L10704. [CrossRef]
44. Ren, H.L.; Jin, F.F. Niño indices for two types of ENSO. *Geophys. Res. Lett.* **2011**, *38*, L04704. [CrossRef]
45. Huang, B.Y.; Thorne, P.W.; Banzon, V.F.; Boyer, T.; Chepurin, G.; Lawrimore, J.H.; Menne, M.J.; Smith, T.M.; Vose, R.S.; Zhang, H.-M. Extended reconstructed sea surface temperature, version 5 (ERSSTv5): Upgrades, validations, and intercomparisons. *J. Clim.* **2017**, *30*, 8179–8205. [CrossRef]
46. Rayner, N.A.; Parker, D.E.; Horton, E.; Folland, C.K.; Alexander, L.V.; Rowell, D.; Kent, E.C.; Kaplan, A. Global analyses of sea surface temperature, sea ice, and night marine air temperature since the late nineteenth century. *J. Geophys. Res. Atmos.* **2003**, *108*, 4407. [CrossRef]
47. Marshall, J.; Adcroft, A.; Hill, C.; Perelman, L.; Heisey, C. A finite-volume, incompressible Navier Stokes model for studies of the ocean on parallel computers. *J. Geophys. Res. Ocean.* **1997**, *102*, 5753–5766. [CrossRef]
48. Kalnay, E. The NCEP/NCAR 40-yr reanalysis project. *Bull. Am. Meteorol. Soc.* **1996**, *77*, 431–477. [CrossRef]
49. Larkin, N.K. On the definition of El Niño and associated seasonal average U.S. weather anomalies. *Geophys. Res. Lett.* **2005**, *32*, L13705. [CrossRef]
50. Mallakpour, I.; Villarini, G. A simulation study to examine the sensitivity of the Pettitt test to detect abrupt changes in mean. *Hydrol. Sci. J.* **2016**, *61*, 245–254. [CrossRef]
51. Das, P.K.; Chandra, S.; Das, D.K.; Midya, S.K.; Paul, A.; Bandyopadhyay, S.; Dadhwal, V.K. Understanding the interactions between meteorological and soil moisture drought over Indian region. *J. Earth Syst. Sci.* **2020**, *129*, 197. [CrossRef]
52. Pyper, B.J.; Peterman, R.M. Comparison of methods to account for autocorrelation in correlation analyses of fish data. *Can. J. Fish. Aquat. Sci.* **1998**, *55*, 2127–2140. [CrossRef]
53. Qu, T. Mixed layer heat balance in the western North Pacific. *J. Geophys. Res. Ocean.* **2003**, *108*. [CrossRef]
54. Bond, N.A.; Cronin, M.F.; Freeland, H.; Mantua, N. *Causes and impacts of the* 2014 warm anomaly in the NE Pacific. *Geophys. Res. Lett.* **2015**, *42*, 3414–3420. [CrossRef]
55. Kara, A.B.; Rochford, P.A.; Hurlburt, H.E. Mixed layer depth variability over the global ocean. *J. Geophys. Res. Ocean.* **2003**, *108*. [CrossRef]
56. de Boyer Montégut, C.; Madec, G.; Fischer, A.S.; Lazar, A.; Iudicone, D. Mixed layer depth over the global ocean: An examination of profile data and a profile-based climatology. *J. Geophys. Res. Ocean.* **2004**, *109*. [CrossRef]
57. Su, J.Z.; Li, T.; Zhang, R.H. The initiation and developing mechanisms of central Pacific El Niños. *J. Clim.* **2014**, *27*, 4473–4485. [CrossRef]
58. Feng, J.; Chen, W.; Tam, C.; Zhou, W. Different impacts of El Niño and El Niño Modoki on China rainfall in the decaying phases. *Int. J. Climatol.* **2011**, *31*, 2091–2101. [CrossRef]
59. Kousky, V.E. The global climate of December 1991–February 1992: Mature-phase warm (ENSO) episode conditions develop. *J. Clim.* **1993**, *6*, 1639–1655. [CrossRef]
60. Iwakiri, T.; Watanabe, M. Multiyear La Niña impact on summer temperature over Japan. *J. Meteorol. Soc. Jpn. Ser. II* **2020**, *98*, 1245–1260. [CrossRef]
61. Woolway, R.I.; Sharma, S.; Weyhenmeyer, G.A.; Debolskiy, A.; Golub, M.; Mercado-Bettín, D.; Perroud, M.; Stepanenko, V.; Tan, Z.; Grant, L. Phenological shifts in lake stratification under climate change. *Nat. Commun.* **2021**, *12*, 2318. [CrossRef]
62. L'Heureux, M.L.; Takahashi, K.; Watkins, A.B.; Barnston, A.G.; Becker, E.J.; Di Liberto, T.E.; Gamble, F.; Gottschalck, J.; Halpert, M.S.; Huang, B. Observing and predicting the 2015/16 El Niño. *Bull. Am. Meteorol. Soc.* **2017**, *98*, 1363–1382. [CrossRef]

atmosphere

MDPI

Article

Hydro-Climatic Variability and Peak Discharge Response in Zarrinehrud River Basin, Iran, Between 1986 and 2018

Farnaz Mohammadi [1], Jaan H. Pu [1,*], Yakun Guo [1], Prashanth Reddy Hanmaiahgari [2], Ozra Mohammadi [3], Mirali Mohammadi [4], Ebrahim Al-Qadami [5] and Mohd Adib Mohammad Razi [5]

1 Faculty of Engineering and Digital Technologies, University of Bradford, Bradford BD7 1DP, UK; f.mohammadi10@bradford.ac.uk (F.M.); y.guo16@bradford.ac.uk (Y.G.)
2 Department of Civil Engineering, Indian Institute of Technology Kharagpur, Kharagpur 721302, India; hpr@civil.iitkgp.ac.in
3 Faculty of Planning and Environmental Sciences, Azad University of Tabriz, Tabriz 5157939396, Iran; mohammadi@zistmohit.ir
4 Department of Civil Engineering, Faculty of Engineering, Urmia University, Urmia P.O. Box 165, Iran; m.mohammadi@urmia.ac.ir
5 Faculty of Civil Engineering and Built Environment, University Tun Hussein Onn Malaysia, Parit Raja, Batu Pahat 86400, Johor, Malaysia; ebrahim@uthm.edu.my (E.A.-Q.); adib@uthm.edu.my (M.A.M.R.)
* Correspondence: j.h.pu1@bradford.ac.uk

Abstract: In recent years, both anthropogenic and climate changes have caused the depletion of surface water resources, shifts in rainfall and accelerations in temperature, which indicates the importance of their examination to flood forecasting analyses. This paper studies the importance of synchronised water management strategies, considering upstream and downstream dynamics using field data from 1986 to 2018. Seasonal and decadal variations show the need for adaptive management strategies to address potential climate change impacts on discharge, precipitation and temperature patterns in the Zarrinehrud River, Iran. The regression analysis was considered via R^2 values, and the statistical analysis was regarded by p-values. The regression analysis of monthly river peak discharge indicates strong correlations between the discharge of specific months (September–October upstream, December–January downstream). By the 2000s and 2020s, both stations showed a shift in peak precipitation to the spring months (April–May for upstream and May–June for downstream). This confirms a synchronisation of rainfall trends, which are influenced by climate changes or regional hydrological patterns. This temporal offset between stations confirms the spatial and seasonal variation in rainfall distribution across the basin. Higher temperatures during the dominant months, particularly late summer to early autumn, accelerate snowmelt from upstream catchments. This aligns with the river discharge peaks observed in the hydrograph. The statistical analysis of river peak discharge indicated that the Weibull (p-value = 0.0901) and the Lognormal (p-value = 0.1736) distributions are the best fitted distributions for the upstream and downstream stations, respectively.

Keywords: flood forecasting; climate change; river peak discharge; precipitation; temperature

1. Introduction

Population growth and climate change play a key role in influencing water resources globally [1]. Climate change has caused intense droughts and flooding events over the years. It has inevitable impacts on the water cycle by altering precipitation patterns and increasing temperatures [2]. In recent years, both anthropogenic and climate changes have caused the depletion of surface water resources, shifts in rainfall and accelerations

 https://doi.org/10.3390/atmos16060681

of temperature, which indicates the importance of their examination to flood forecasting analyses [3,4]. There are two strategies to cope with flood hazards: (i) non-structural methods, including real-time flood forecasting with numerical methods to assist decision makers, and (ii) structural methods by building protection measures like dikes and dams [5]. Numerous studies have investigated the effects of climatic factors on river behaviour, particularly in flood-prone basins. Lake Urmia, the world's second-largest salt lake, located in the northwest of Iran, encounters too many environmental challenges. Addressing the escalating impacts of climate change and human activities while considering the construction of dams along the rivers that flow into the lake, which are entwined in domestic and agricultural developments, has caused an increase in water demands in recent decades. The water level of the lake has plunged into a great decrease of 90% in a 15-year span from 2000 to 2015 [6,7]. Due to the continuous degradation of the water level and droughts at the Lake Urmia Basin (LUB), studying one of its prominent rivers, the Zarrinehrud River, would have profound significance in comprehending its future environmental conditions. The hydrometric stations used in this study are located at a significant distance from the dam structures to reduce the direct operational impact on the discharge readings, thereby allowing for a more representative analysis of the river's downstream behaviour. Many researchers have considered the river peak discharge, precipitation and temperature for flood forecasting to reduce the uncertainties in hydrological and hydraulics modelling for various rivers [8–11].

Kumar et al. [12] described sediment flow transportation parameters using machine-learning techniques to look at sediment transportation. The distribution of suspended sediment over the characteristic height inside the flow was compared with other experimental data in the same way as it is investigated on data in this study to perform a comparison. Therefore, it is essential to conduct data analyses before any further examination of data, like historical data modelling. There are several flood forecasting methods, which introduce different data types. Kazemi Garajeh et al. [2] studied the increases in air temperature (AT) and evapotranspiration (ET) in the LUB, Iran, between 2000 and 2021. A decline in annual precipitation occurred, as evidenced by the negative correlation between AT, ET and water surface area. The findings of this study indicated correlation coefficients of −0.55, −0.59 and 0.39 between the water area, AT, ET and precipitation, respectively. According to the results in their study, climate change has affected water resources from 1984 to 2021 in the LUB, as the same years are partly involved in this study. Asadi et al. [6] employed the following daily hydrological alterations: precipitation, temperature, streamflow and land use for 24 years during the years 1995–2019 to investigate the implementation of environmental flow in a water allocation model and analyse water demand reliabilities at the Zarrinehrud River Basin, Iran, using SWAT+, WEAP and IAHRIS. They emphasised the changes to the hydrological regime caused by the Bukan dam. Their research covers the same years of this study and also explains the crucial role of the Bukan dam while investigating months with river peak discharge.

Toth et al. [13] considered past rainfall depth as a sole input to compare the accuracy of the short-term rainfall forecasts obtained with time-series analysis techniques, including linear stochastic auto-regressive moving average (ARMA) models, artificial neural networks (ANN) and the non-parametric nearest-neighbours method for forecasting storm rainfalls that occurred in the Sieve River basin, Italy, in the period of 1992–1996. The overall results show that even if the rainfall forecasting performance measures indicate only a weak-to-moderate relationship between the forecasted and observed values, considering time-series analysis techniques provides an improvement in flood forecasting accuracy. It was concluded that past rainfalls alone are not sufficient to make precipitation projections. Therefore, precipitation, temperature and river peak discharge would be more suitable inputs for flood forecasting, and the historical data are analysed in this paper. Te Linde et al. [14] explored flood management for the year 2050 along the Rhine River basin, as a cross-boundary river located in northwest Europe, which was determined between the upstream and downstream. It was planned to simulate long discharge series (10,000 years) to estimate peak discharge with a low probability of flooding events using flood waves at four locations with four return periods. They indicated that upstream flooding has a profound decreasing effect on the simulated peak water levels and discharges along the main Rhine branch and downstream in the Netherlands, where there are two hydrometric stations located upstream and downstream of the river. These stations were selected to perform flood management in this study as well. Supriya et al. [15] investigated the regression analysis at the Vellar river basin in South India by annual maximum daily rainfall, stream flow and area for each catchment. A Multiple Regression Model (MLR) was used to understand the relationship between the weighted annual maximum daily rainfall, catchment area and annual maximum daily stream flow. It was found that a regression analysis is an effective statistical method for flood forecasting techniques based on catchment characteristics, and this technique is the basis of this paper's discussion.

Shirmohammadi et al. [16] studied the effects of climate change on stream flow variation in the LUB, Iran, by using daily statistics, including precipitation, minimum and maximum temperature and runoff during the base period (1981–2010). The performance analysis of the observed and simulated data indicated that the LUB faces decreasing stream flows, declining precipitation and rising temperatures in the future. Their findings confirmed the importance of data analysis to create scenarios to generate projections. Behboudian et al. [17] assessed the impacts of climate change using daily precipitation and minimum–maximum air temperature for the period of 2019–2049 in the Zarrinehrud River Basin, Iran. To simulate various climate change scenarios, a SWAT model was used to simulate the rainfall-runoff processes and evaluate land management practices from two hydrometric stations: upstream (33023) and downstream (33917). They provided the time series of the monthly inflow to the reservoir from 2019 to 2049. They concluded that the priority order of the water demand supply is municipal, industrial, environmental and agricultural, with the drinking water demand projected to increase by 50%. Additionally, drought resilience was found to be as equally important as flood management in this region. Kazemi Garajeh et al. [18] explored the anthropogenic activities on surface water resources across the LUB, Iran. Air temperature, cropland, potential evapotranspiration, snow cover, precipitation, built-up areas and groundwater salinity data acquired from 2000 to 2021 were considered and imported using implemented code on the Google Earth Engine (GEE) platform. The Mann–Kendall (MK) non-parametric trend test and Pearson correlation analysis were conducted for the evaluation of the results. Climatic factors such as air temperature, potential evapotranspiration, precipitation and snow cover demonstrate negative correlations of −0.56 and −0.09 for air temperature and snow cover, with surface water resources being identified as the main elements to employ for flood forecasting. Kumar et al. [19] investigated geomorphological alterations due to climate change and anthropogenic causes in the Gandak River, which is located in Nepal and merges with the Ganga River in India. They analysed river migration from 1989 to 2022. It was observed that erosion and river shifting are progressing at many locations, and many villages were wiped off during the floods. There are several similarities with these papers and this study that can be indicated: (i) selecting an upstream and downstream hydrometric station as a benchmark to perform comparisons, (ii) more than 30 years of available data and (iii) employing river peak discharge and climatic data like precipitation and air temperature, predominantly for flood forecasting purposes.

The objective of this study is to use river peak discharge data as the inflow data while using precipitation and temperature data as the climatic data to understand their trends during recent decades and assess how the weather paradigm affects the river's behaviour. In this study, 33 years of historical data (from 1986 to 2018) for the Zarrinehrud River, Iran, located in one of the most vulnerable basins in the world, the LUB, is considered to explore approaches for effective flood management. To minimise the impact of dams on the investigated data, a sufficient distance from those dams to the measuring stations was kept. Moreover, the potential influence of dam regulation has been thoroughly acknowledged and fully discussed in the Results and Discussions Section to provide clarity on the interpretation of the observed discharge patterns. Unlike studies that focus primarily on projections, this paper focuses on examining the historical records to understand past behaviours, identify key influencing factors such as climatic changes and dam impacts, and determine the dominant months that are prone to potential flooding events for better preparedness in future modelling efforts.

2. Materials and Methods

2.1. Study Area

This case study is located in the northwest of Iran and encompasses the Lake Urmia Basin (LUB), a crucial geographical and ecological region covering a land area of 51,876 km^2. LU occupies 10% of the area, which is depicted in Figure 1 [16]. This terminal relies on water inflow from 17 permanent and 12 seasonal rivers, and the main part of this river is located in Shahin Dej county, Iran. The case study highlights the significance of this region in terms of its environmental and hydrological characteristics [16]. In particular, the Zarrinehrud River, serving as one of the primary inflows among those rivers, contributes 41% of the surface water flow into Lake Urmia [17]. Lake Urmia, encompassing an expanse of 6000 km^2, is the world's second-largest hypersaline lake, where the Zarrinehrud River is located at the lowest point of the LUB [18]. Over the years, the lake's size has been greatly reduced due to multiple factors, including excessive agricultural water usage, river dams and droughts from climate change impacts [19]. The LUB region is considered a semi-arid and dry-cold climate with an average annual rainfall of 359 mm, a minimum temperature of 4.3 °C and a maximum temperature of 17.7 °C [20]. The evaporation rates show a range from 930 to 1513 mm per year [21]. The Zarrinehrud River basin, with a land area of 12,000 km^2, extends for 320 km. The region comprises semi-cold-wet or cold-wet conditions in mountainous areas and semi-dry conditions near Lake Urmia. The rainfall within the basin varies between 200–800 mm/yr [20,21]. Lake Urmia's inflow comprises 25% from direct precipitation and 75% from 60 permanent or seasonal rivers [22], which makes it important to study one of them.

The Zarrinehrud River's hydrometric stations are going to be introduced as the upstream-33023 and downstream-33917 stations throughout this study and are shown in Figure 2. The upstream-33023 station is located 5 kilometres down from the Bukan Dam, and the downstream-33917 station, as the ultimate station of the Zarrinehrud River basin, is located after the Nowruzlu dam. The Nowruzlu dam is located 70 km downstream of the Bukan Dam. At this dam, the right and left bank canals are used to supply drinking water to the city of Tabriz and the surrounding areas, respectively. Between the Bukan Dam and the Nowruzlu Dam, eight seasonal rivers named Qoreh Chai, Holasu, Hachasu, Aghche, Joshato, Sourin, Ajarlu and Qoreh Chai connect to the Zarrinehrud River. The Nowruzlu diversion dam was built on the Zarrinehrud River to control floods in this river and for agricultural and industrial purposes [23,24]. In Figure 2, the locations of Lake Urmia and the Zarrinehrud River are imported from Google Maps, and the locations of the Bukan dam, Upstream-33023 station, Nowruzlu dam and downstream-33917 station are illustrated.

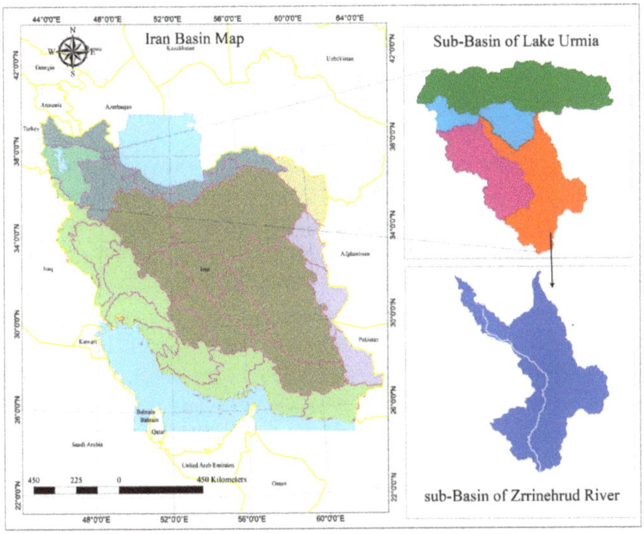

Figure 1. Case study—Zarrineh-Rood River, West Azerbaijan, Iran.

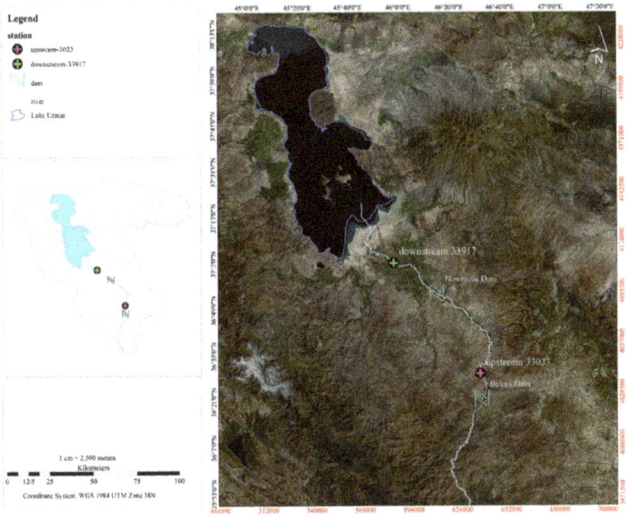

Figure 2. Location of hydro-climatic stations and dams along the river.

2.2. Dataset

Herein, four datasets, including annual river peak discharge, monthly river peak discharge, monthly precipitation, and monthly temperature of the upstream-33023 and downstream-33917 stations, are employed for the years between 1986 and 2018. These datasets were acquired from the Regional Water Organisation of West Azarbaijan, Iran. Two methodological challenges exist for hydrological data analyses in the Zarrinehrud River: (i) there is a short period of data available to conduct surveys, which would lead to some uncertainties and analysis pulses, and (ii) in the last decades, both the climatic patterns and the river basin have shifted. This paper will use the thirty-three years of data spanning from 1986 to 2018, which contains the 1990s, 2000s and 2010s, to examine the potential components for flood forecasting through the Zarrinehrud River. The reason for selecting the upstream-33023 and downstream-33917 stations was due to the existence of two important dams: the Bukan dam at the upstream and Nowruzlu dam at the downstream, which are noteworthy to analyse the river flow's behaviour and investigate the affected data throughout the years in this region.

2.3. Data Analysis

In this paper, the calibration process began with analysing the hydrographs of the upstream-33023 and downstream-33917 stations. The hydrographs were used to explore the behaviour of the water inflow (i.e., extremes or droughts) along the river between both the upstream and downstream stations. To address projections, the historical river peak discharges were examined to extract dominant months to understand the data trends. The validation of the data was conducted via regression analysis. Climate change led to evolving flood management in this region. To overcome the potential changes, two stations were selected to perform the comparison and conduct a statistical exploration of regional data as a benchmark, and the target was to illustrate the trends of the annual river peak discharge hydrographs, monthly river peak discharge, monthly precipitation and monthly temperature, and study the correlation between those historical data to discuss their interaction effects throughout the years for further simulation processes. This study aims to initially analyse the observed data and distinguish between dominant months to create scenarios and make projections.

Validation of Data

To study flooding on the Zarrinehrud River, Iran, the river peak discharge (m^3/sec) was requested for an available period as dependent data, and accordingly, the independent data, including precipitation (mm) and temperature (°C), were employed. The real-time data experienced a preparatory phase in this study, allowing us to understand the river's behaviour without any manipulation before delving into modelling. Due to this, the regression analysis was considered via R^2 data values, and the statistical analysis was regarded by p-values on the real monthly river peak discharge as a dependent variable, including Normal, Lognormal, Gamma and Weibull, which were extracted for the upstream and downstream stations to check the skewness of the data and complete an initial visual fitting. Meanwhile, the goodness of fit tests, the Kolmogorov–Smirnov (K–S) test and Anderson–Darling (A–D) test, and the trend analysis tests, the Mann–Kendall test and Sen's Slope Estimator, were applied considering p-values. Those statistical tests are all indicated in Table 1.

Table 1. The statistical tests.

Kolmogorov–Smirnov (K–S) Test	Good for general fitting. Works best with continuous distributions like Normal, Lognormal, Weibull, etc.
Anderson–Darling (A–D) Test	Excellent for flood peak data, because extreme values matter a lot.
Mann–Kendall	Used for trend detection in time-series data (discharge over time)
Sen's Slope Estimator	Calculates the magnitude of the trend detected by Mann–Kendall

3. Results and Discussion

Scientists have highlighted the necessity of collecting more accurate data to reduce the uncertainties in hydrological and hydraulics analyses. For practical flood forecasting, the first step is to identify data and its time step and investigate the river flow's behaviour like in [2,4,6]. The objective of this paper is to study the capability of the data, particularly the accuracy of its representative factors and the way climate change affects flood management. For flood forecasting in the Zarrinehrud River, Iran, both the upstream-33023 and downstream-33917 stations play a key role. The annual river peak discharge, monthly river peak discharge, monthly precipitation and monthly temperature of those stations in recent decades were analysed spanning from 1986 to 2018 in further sections. The upstream-33023 station shows more direct control by the Bukan Dam, leading to more regulated discharge patterns, which aligns with the findings of [6], and on the other hand, the downstream-33917 station has higher variability due to the contributions from the seasonal rivers and the influence of the Nowruzlu Dam.

3.1. *Annual River Peak Discharge Hydrographs*

Figure 3 illustrates the annual river flow rate for each hydrometric station. The river peak discharges for both stations have an almost identical range, considering their geographical features. The flow pattern in Figure 3 shows the fluctuations of the upstream-33023 station over the years, with notable peaks around 1996, with the value of 690 m^3/s. This station is influenced by releases from the Bukan Dam, suggesting controlled regulation of water flow. Given the significant distance of the measurement stations from the dams, the impact of dam operations on discharge readings is minimised. While detailed operational records were unavailable, the influence of dam regulation has been thoroughly discussed to support the validity of the statistical characterisation of the managed flow regime. Due to this, the variability in discharge suggests the dam's operation plays a critical role in regulating the upstream flow [6]. Considering the downstream-33917 station's graph, it shows similar peaks in the discharge, but the magnitude and timing of the peaks slightly differ compared to the upstream-33023 station. This could be due to the presence of seasonal rivers and contributions from the tributaries between the dams, adding complexity to the flow, and the hydrograph also highlights the downstream-33023 station's role in flood control and its potential to manage seasonal variability. It is concluded that at the upstream-33023 station, the variability in the peak discharge shows that the Bukan Dam likely moderates seasonal flows to reduce floods during wet seasons and supplement water during dry periods. Some years exhibit very low or almost negligible peaks (e.g., mid-2000s). This may be indicative of drought conditions or operational constraints at the dam.

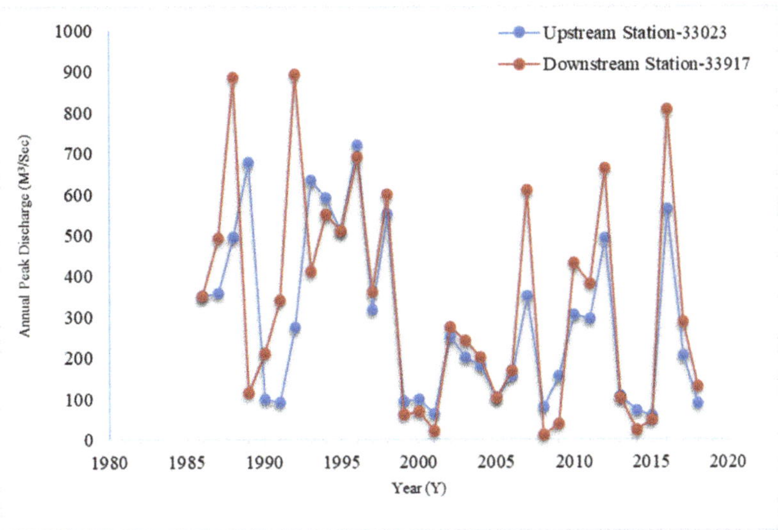

Figure 3. Hydrographs of upstream-33023 and downstream-33917 stations of Zarrinehrud River.

The highest peaks (e.g., early 1990s and 2015) align with potential periods of heavy rainfall or dam releases for flood control or irrigation demand. It is notable that the hydrograph of the downstream-33917 station demonstrates higher variability than the upstream-33023 station, especially during wet years. Seasonal rivers likely contribute to sudden spikes in flow during the rainy season (spring or fall), amplifying the downstream discharge. Peaks like those in 1995, 2015 and 2018 appear to correspond with upstream peaks, but with added contribution from seasonal rivers. Particularly, the alignment or misalignment of peak years between the stations shows how tributaries and dams interact to influence the flow behaviour. Considering the long-term trends, there is a decreasing peak discharge in the 2000s, which could be attributed to reduced rainfall (climatic variability), increased water withdrawals for agriculture and urban demands or dam operations prioritising water storage over flood control. In the years 2007, 2012 and 2016, both hydrographs exhibit a jump to higher peaks, especially downstream, with the values of 610, 661.42 and 807, respectively. This could suggest recovery from drought conditions, higher rainfall or snowmelt, or operational changes in dam release schedules. On the other hand, during high-peak years (e.g., 1990–1995 and 2015–2018), the downstream discharges exceed the upstream peaks, illustrating the impact of seasonal rivers and catchment area dynamics.

Both hydrographs show that the Bukan Dam plays a dominant role in moderating the upstream flow. During specific years like 1993 and 1999, the dam retains significant volumes of water to mitigate floods or store water for dry seasons. The Nowruzlu Dam and its associated diversion infrastructure, located downstream, manage flood events effectively, as shown by the limited extreme fluctuations in downstream peaks. The seasonal rivers connecting downstream to the Zarrinehrud River likely exacerbate peak discharges during heavy rain or snowmelt events. Urbanisation, agriculture and canal systems may also contribute to reduced baseflows and increased variability in the downstream flow. Overall, in specific years, there seems to be synchronisation between the upstream and downstream peaks (e.g., 1995 and 2015), indicating the rapid transmission of floodwaters. Meanwhile, in other cases, the downstream peaks lag in comparison to the upstream peaks due to the buffering effect of the Nowruzlu Dam and seasonal river inflows. In Figure 3, the river peak discharges seen upstream are attenuated downstream, which is related to the combined effects of the dams and tributaries, and the river's natural storage capacity.

The regression analysis predicts the magnitude of flood events considering various recurrence intervals for most streams [25]. This is also analysed in this study in the Results Section, where the statistical equations are extracted.

3.2. Monthly River Peak Discharge

The hydrological behaviour of the Zarrinehrud River has been analysed using the monthly river peak discharge data of those two key stations in Figures 4 and 5. The results of the regression analysis are summarised in Table 1, with further insights drawn from the hydrographs.

Table 2 indicates that the upstream-33023 station, located 5 km downstream of the Bukan Dam, exhibits peak discharges predominantly in September–October, during early autumn. The statistical relationship between the discharge and the independent variable is captured by a polynomial equation. The R^2 value of 0.786 for the upstream-33023 station implies a good fit. The peak discharge behaviour of the upstream-33023 station is likely influenced by controlled releases from the Bukan Dam, which aligns with agricultural water demands and the reduction in reservoir levels post-summer. The sharper fluctuations observed in the upstream hydrograph reflect this controlled response to operational priorities, with limited natural contributions from tributaries. The downstream-33917 station, located 70 km downstream of the Bukan Dam and influenced by the Nowruzlu Dam and eight seasonal tributaries, exhibits a distinct seasonal pattern, with peak discharges occurring in December–January during the winter months. This is indicated by a power equation that achieves a higher R^2 value of 0.845, indicating a stronger relationship compared to the upstream station in the time period between the years 1986 and 2018.

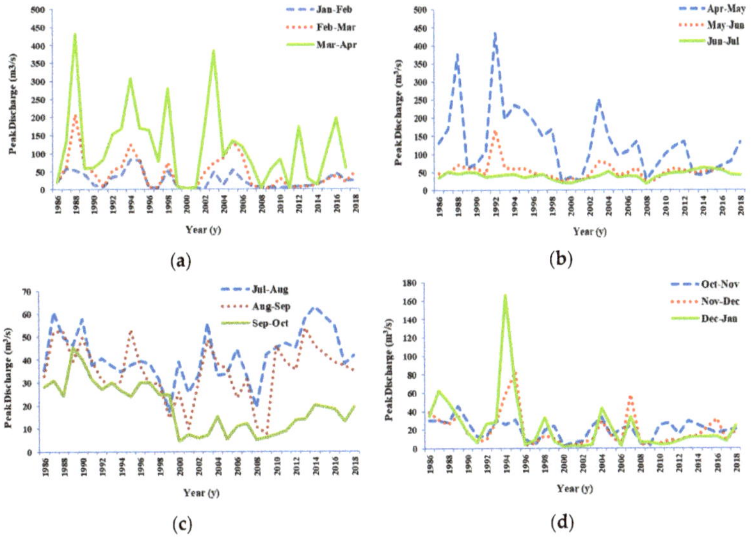

Figure 4. Seasonal peak discharge at the upstream-33023 station: (**a**) winter, (**b**) spring, (**c**) summer and (**d**) autumn.

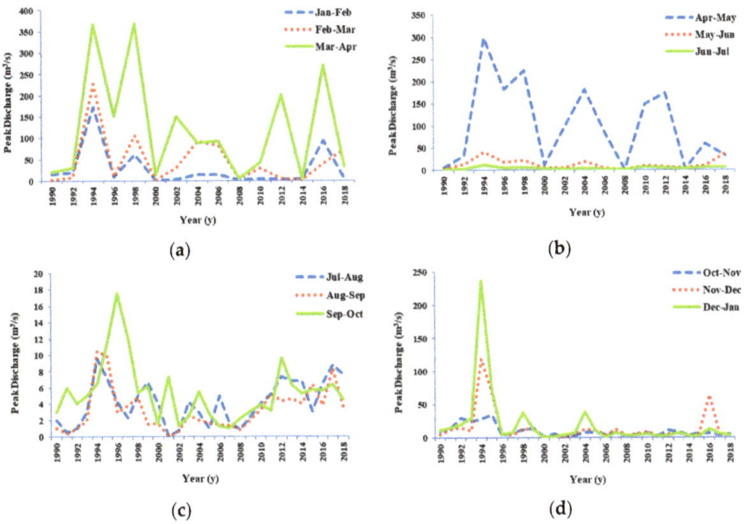

Figure 5. Seasonal peak discharge at the downstream-33917 station: (**a**) winter, (**b**) spring, (**c**) summer and (**d**) autumn.

Table 2. Regression analysis of monthly river peak discharge.

Stations	R^2	Statistical Equation	Months with the Highest R^2
Upstream-33023	0.786	$y = 0.0574x^2 - 2.6859x + 42.976$	September–October
Downstream-33917	0.845	$y = 60.39x^{-0.967}$	December–January

3.3. Monthly Precipitation

The temporal evolution of the precipitation at the upstream-33023 station is presented in Figures 6–8. The precipitation at the upstream-33023 station peaks in September–October during the early autumn in the 1990s and peaks in April–May during the early spring in the 2000s and 2010s. The R^2 values in Table 3 for the regression analysis show a progression over three decades. In the 1990s, the power equation shows a lower predictive accuracy of $R^2 = 0.661$. However, the R^2 values for other decades, i.e., the 2000s and 2010s, show a moderate predictive strength and a high predictive reliability, which are 0.719 and 0.829, respectively. The hydrograph peaks in the spring align with the April–May precipitation trends in the 2000s and 2010s, indicating a strong linkage between precipitation events and river discharges. The increasing R^2 values over those decades suggest the enhanced predictability of peak discharges, which are potentially influenced by anthropogenic factors such as the regulation of water flows by the Bukan dam at the upstream.

In Figures 9–11, the precipitation trends are demonstrated for three decades at the downstream-33917 station. Due to those graphs, the R^2 values in Table 4 for the regression analysis describe varying degrees of correlation. In the 1990s, the $R^2 = 0.435$ with a polynomial equation, indicating weaker predictive reliability during March–April. In the 2000s, the precipitation data shows a perfect fit for May–June, and in the 2010s, it maintains high predictive reliability during the same period of May–June. The downstream hydrograph peaks during May–June align with the observed precipitation trends, highlighting precipitation as the primary driver of the downstream discharges. The exact fit in the 2000s reflects the regulation of downstream flows by the Nowruzlu Dam, which minimises variability and aligns the discharge patterns with precipitation events.

The upstream station experienced its peak precipitation earlier in the year during the 1990s (September–October), while the downstream station's peaks occurred in March–April. This suggests a potential lag in water movement or differing rainfall distribution across the basin. By the 2000s and 2010s, both stations show a shift in peak precipitation to spring months (April–May for upstream and May–June for downstream). This indicates a synchronisation of rainfall trends, possibly influenced by climate change or regional hydrological patterns.

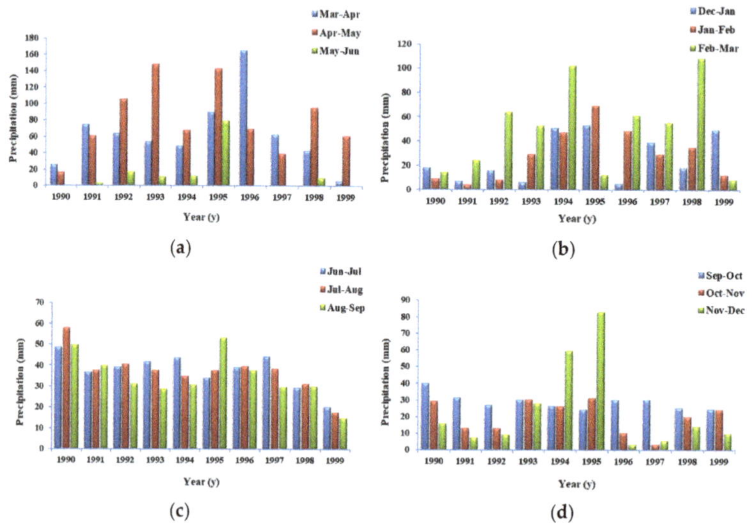

Figure 6. Precipitation at upstream-33023 station in 1990s: (**a**) winter, (**b**) spring, (**c**) summer (**d**) autumn.

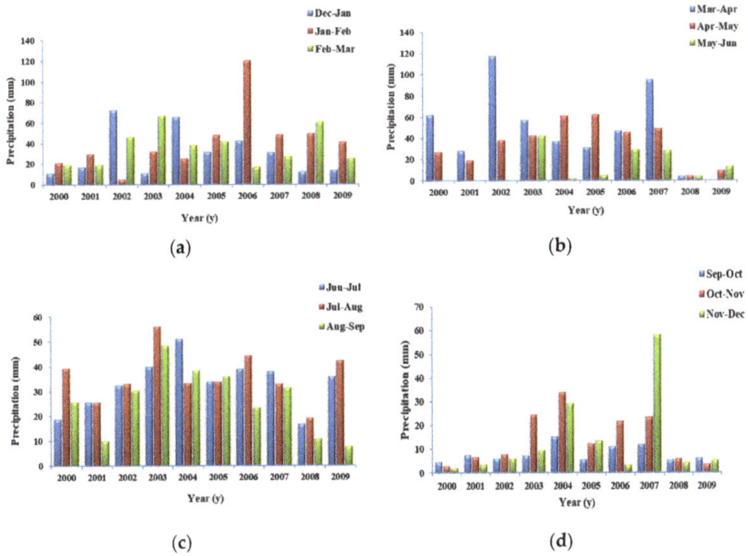

Figure 7. Precipitation at upstream-33023 station in 2000s: (**a**) winter, (**b**) spring, (**c**) summer (**d**) autumn.

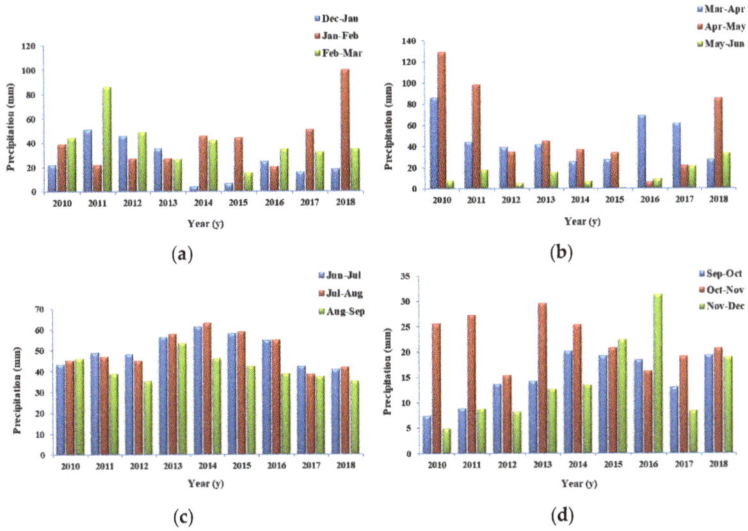

Figure 8. Precipitation at upstream-33023 station in 2010s: (**a**) winter, (**b**) spring, (**c**) summer (**d**) autumn.

Table 3. Regression analysis of monthly precipitation for upstream-33023 station.

Stations	R^2	Statistical Equation	Months with the Highest R^2
Upstream-33023	0.661	$y = 36.487x^{-0.163}$	September–October 1990s
	0.719	$y = -2.2008x^2 + 23.011x - 6.2833$	April–May 2000s
	0.829	$y = 4.7816x^2 - 55.719x + 181.69$	April–May 2010s

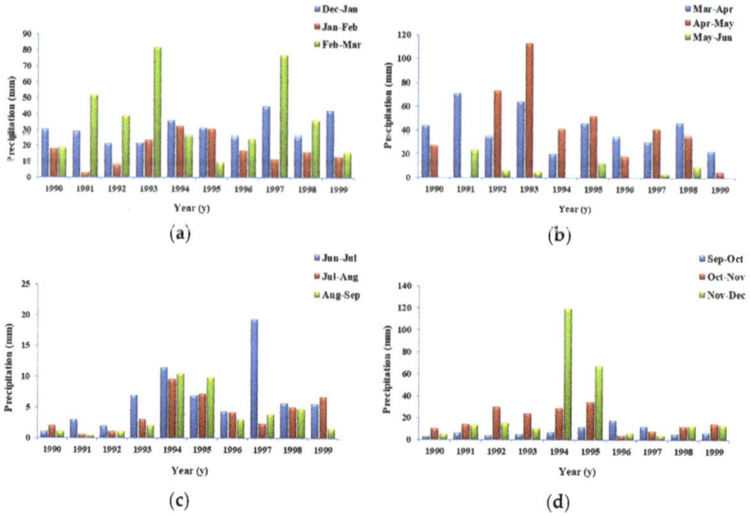

Figure 9. Precipitation at downstream-33917 station in 1900s: (**a**) winter, (**b**) spring, (**c**) summer and (**d**) autumn.

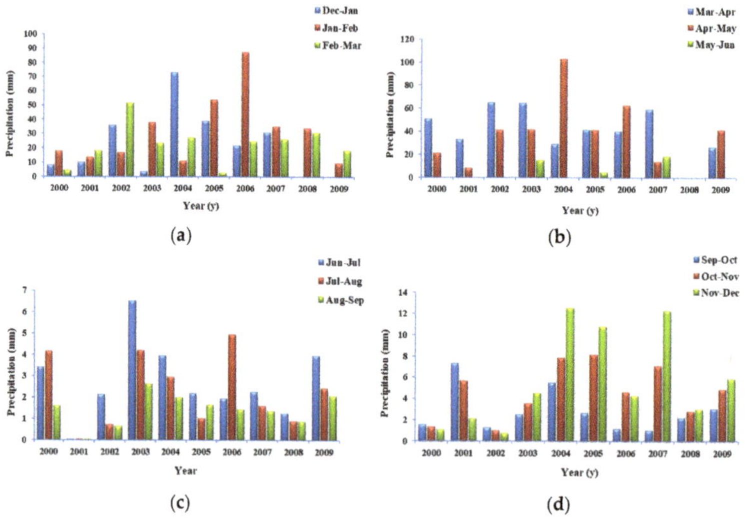

Figure 10. Precipitation at downstream-33917 station in 2000s: (**a**) winter, (**b**) spring, (**c**) summer (**d**) autumn.

Table 4. Regression analysis of monthly precipitation for downstream-33917 station.

Stations	R^2	Statistical Equation	Months with the Highest R^2
Downstream-33917	0.435	$y = -1.9089x^2 + 16.424x + 29.313$	March–April 1990s
	1	$y = 3.0625x^2 - 35.875x + 109.5$	May–June 2000s
	0.921	$y = 1.4792x^2 - 17.017x + 46.438$	May–June 2010s

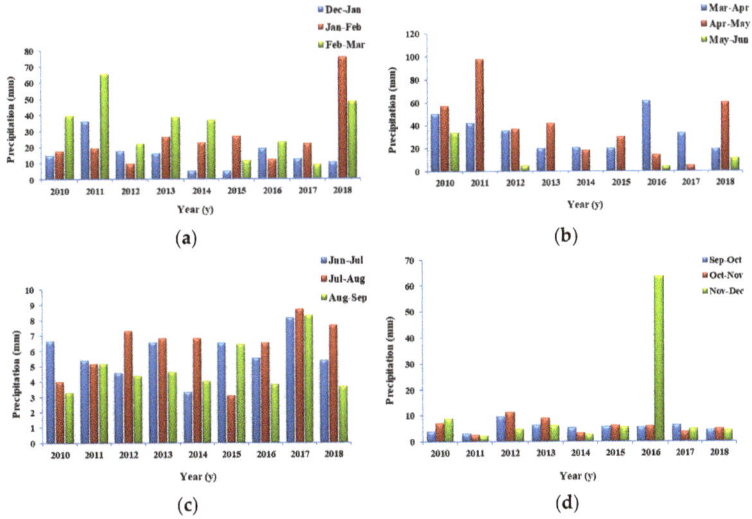

Figure 11. Precipitation at downstream-33917 station in 2010s: (**a**) winter, (**b**) spring, (**c**) summer (**d**) autumn.

3.4. Monthly Temperature

Figures 12–17 represent three recent decades of seasonal temperature analyses that were categorised as the 1990s, 2000s and 2010s for each station separately. Those graphs comprise seasonal monthly temperature data for both the upstream-33023 and downstream-33917 stations. Table 5 shows the regression analysis of the monthly temperature for the upstream-33023 station over three decades. The table highlights the dominant months for temperature peaks and their potential impact on the hydrology of the Zarrinehrud River.

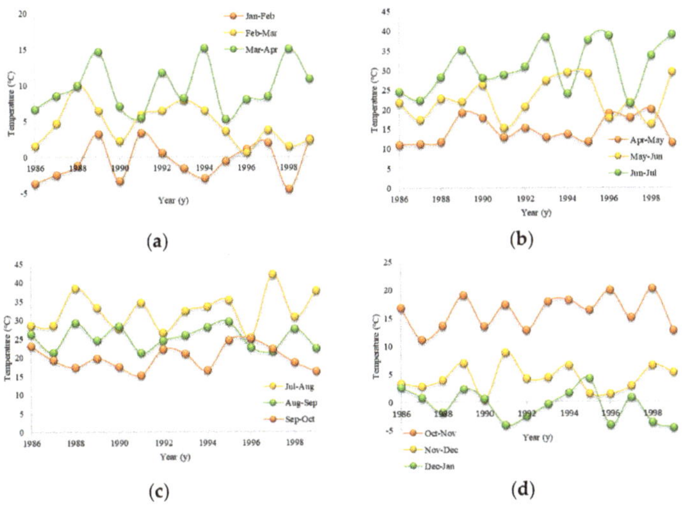

Figure 12. Temperature at the upstream-33023 in 1990s, (**a**) Winter, (**b**) Spring, (**c**) Summer, (**d**) Autumn.

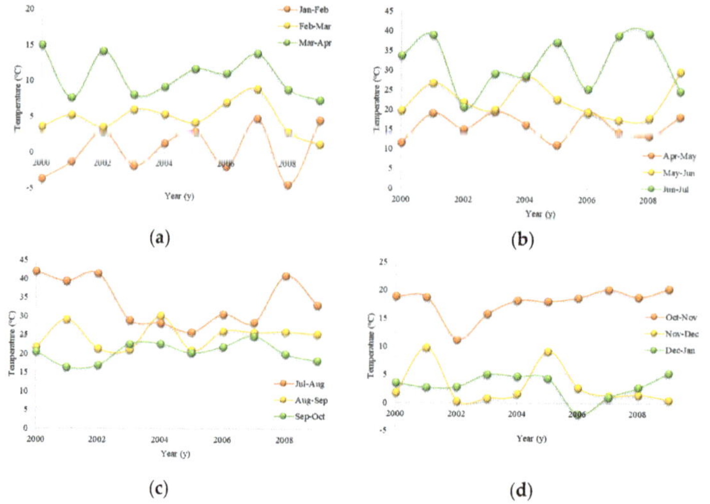

Figure 13. Monthly temperature at the upstream-33023 in 2000s, (**a**) Winter, (**b**) Spring, (**c**) Summer, (**d**) Autumn.

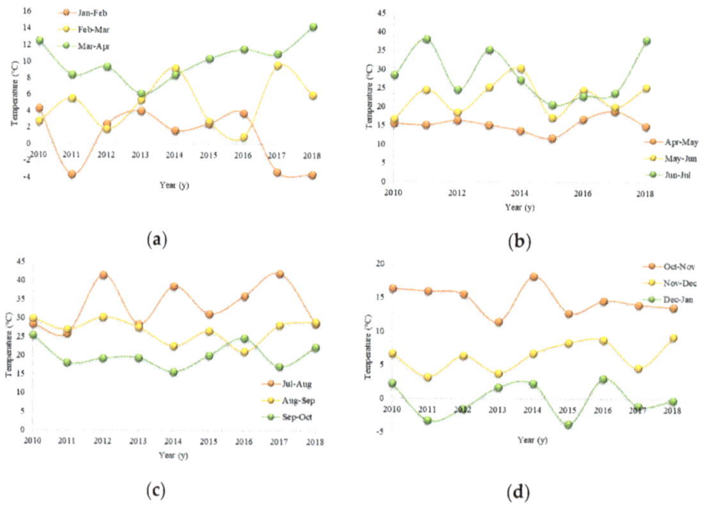

Figure 14. Monthly temperature at the upstream-33023 in 2010s, (**a**) Winter, (**b**) Spring, (**c**) Summer, (**d**) Autumn.

The temperature trends over the years include the following: (a) 1990s: The highest $R^2 = 0.818$ indicates that the monthly temperature variability during this decade was well-explained by a polynomial equation. The dominant temperature months spanned from September to October, confirming the steady influence of temperature on annual hydrology. (b) 2000s: A lower $R^2 = 0.668$ suggests reduced consistency in the temperature–discharge relationship. The temperature peaks in November–December, reflecting a shift in seasonal patterns or changes in climatic conditions. (c) 2010s: The $R^2 = 0.748$ shows a stronger relationship between temperature and hydrology compared to the 2000s. The dominant months (September–October) suggest that the temperature effects remained impactful throughout the year. The changes in coefficients over decades highlight potential shifts in climatic patterns influencing hydrological processes. The monthly hydrograph for the upstream station shows annual river peak discharges in September–October, coinciding with the dominant temperature months in the 1990s and 2010s. Higher temperatures during the dominant months, particularly in the late summer to early autumn, accelerate snowmelt from upstream catchments. This aligns with the discharge peaks observed in the hydrograph.

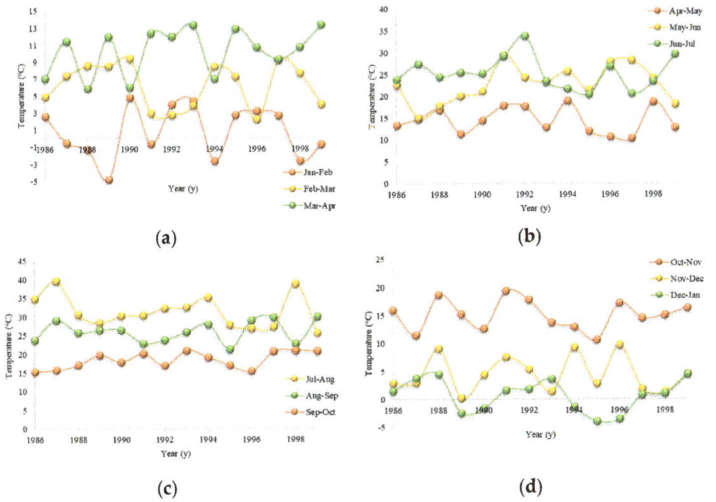

Figure 15. Monthly temperature at the downstream-33917 in 1990s, (**a**) Winter, (**b**) Spring, (**c**) Summer, (**d**) Autumn.

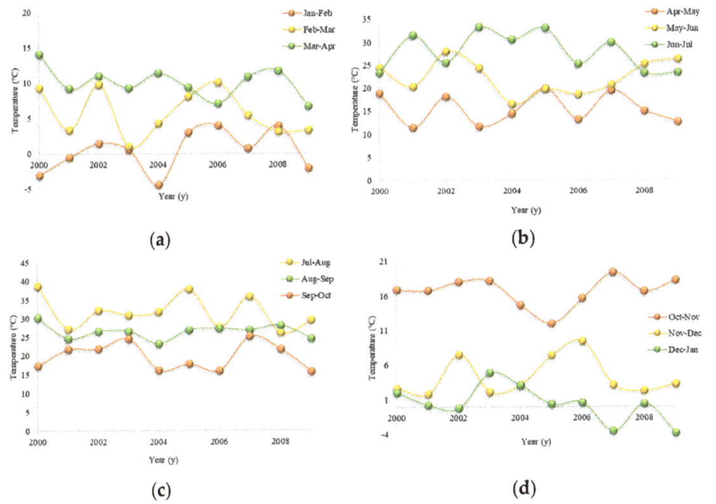

Figure 16. Monthly temperature at the downstream-33917 in 2000s, (**a**) Winter, (**b**) Spring, (**c**) Summer, (**d**) Autumn.

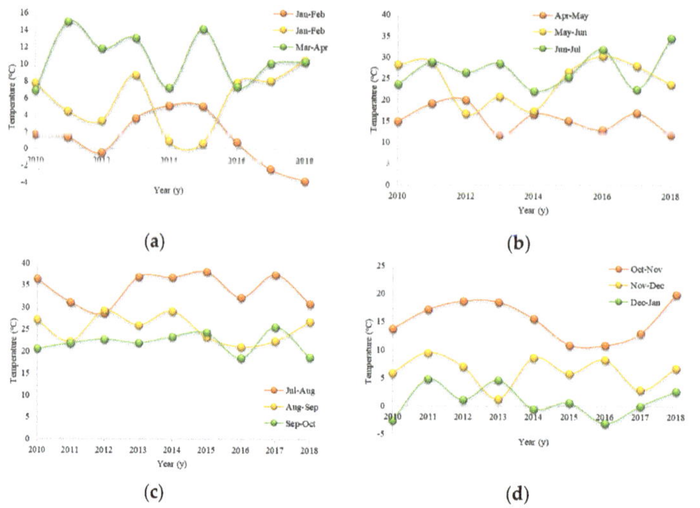

Figure 17. Monthly temperature at the downstream-33917 in 2010s, (**a**) Winter, (**b**) Spring, (**c**) Summer, (**d**) Autumn.

Table 5. The regression analysis of the monthly temperatures for the upstream-33023 station.

Stations	R^2	Statistical Equation	Months with the Highest R^2
Upstream-33023	0.818	$y = 0.4475x^2 - 1786.7x + 2E{+}06$	September–October 1990s
	0.668	$y = -1.0428x^2 + 4181.7x - 4E{+}06$	November–December 2000s
	0.784	$y = -0.2416x^2 + 974.75x - 983074$	September–October 2010s

Table 6 provides the regression analysis of the monthly temperatures for the downstream-33917 station across three decades. The table includes the statistical equations, the R^2 values and the dominant months of temperature variation for each decade. Temperature trends over time include the following: (a) In the 1990s: The highest $R^2 = 0.839$ indicates a strong fit of the quadratic equation, with temperature peaks in July–August, suggesting the influence of summer heat. (b) In the 2000s: A lower $R^2 = 0.567$ points to the weaker predictability of the temperature–discharge relationship. The peaks shifted to April–May, reflecting an earlier onset of high temperatures, which is linked to climate variability. (c) In the 2010s: The $R^2 = 0.779$ shows an improved correlation compared to the 2000s. The peak months remained April–May, indicating a continued trend of springtime temperature dominance. The shift from July–August (1990s) to April–May (2000s and 2010s) indicates earlier warming trends in the basin, which can be attributed to the changing climatic conditions and seasonality in the region. The statistical equations highlight the nonlinear temperature variations over decades, with the coefficients reflecting changes in both temperature magnitude and seasonality.

Table 6. The regression analysis of the monthly temperature for the downstream-33917 station.

Stations	R^2	Statistical Equation	Months with the Highest R^2
Downstream-33917	0.839	$y = -1.8727x^2 + 7456.1x - 7E{+}06$	July–August 1990s
	0.567	$y = -3.8468x^2 + 15426x - 2E{+}07$	April–May 2000s
	0.779	$y = 2.6058x^2 - 10491x + 1E{+}07$	April–May 2010s

3.5. Statistical Distribution Fitting and Goodness of Fit Test

For reliable analyses in this study, several statistical distribution fittings, including Normal, Lognormal, Gamma and Weibull, are applied to conduct statistical tests, like the Kolmogorov–Smirnov (K–S) test, Anderson–Darling (A–D) test, Mann–Kendall test and Sen's Slope Estimator, on the historical river peak discharge data as dependent data for both the upstream and downstream stations. To evaluate the suitability of the candidate distributions for modelling the discharge data, the Kolmogorov–Smirnov (K–S) test is summarised in Table 7, including K–S statistics and corresponding *p*-values for each fitted distribution. Ref. [26] declared that hydrologic variables, including streamflow, are routinely found to possess heavy tails. Therefore, hydrological data, particularly river peak discharges, often exhibit characteristics such as heavy tails and skewness, which can lead to significant deviations from standard theoretical distributions relating to low *p*-values when conducting goodness of fit tests, as can be seen in Table 7. A smaller K–S statistic indicates a better fit. At the upstream station, the K–S statistics are comparatively lower for all distributions. Notably, the Weibull (0.0901) and Lognormal (0.0962) distributions exhibit the smallest K–S values, suggesting a better approximation of the upstream discharge behaviour relative to the downstream station. At the downstream station, the K–S statistics are generally higher, particularly for the Gamma (0.9313) and Normal (0.4241) distributions, indicating a poorer fit compared to the upstream station. The Lognormal distribution (smallest K–S statistic of 0.1736) outperformed among all distributions in the downstream station.

Table 7. The statistical Kolmogorov–Smirnov test for the monthly river peak discharge for both stations.

Location	Distribution	K–S Statistic	*p*-Value
Upstream-33023	Normal	0.2316	0.0000
	Lognormal	0.0962	0.0012
	Gamma	0.1027	0.0004
	Weibull	0.0901	0.0030
Downstream-33917	Normal	0.4241	0.0000
	Lognormal	0.1736	0.0000
	Gamma	0.9313	0.0000
	Weibull	0.3725	0.0000
	Normal	0.4241	0.0000

For visual support, the Anderson–Darling (A–D), Mann–Kendall and Sen's Slope Estimator tests are shown in Figures 18 and 19. The K–S test indicates that discharge at the upstream station is more statistically regular and fits classical distributions better than the discharge from the downstream station, which shows greater irregularity and extreme values. The A–D test suggests that the downstream site displays stronger deviations from normal behaviour, pointing to a more skewed and extreme discharge distribution. According to the Mann–Kendall test trend, the discharge at the upstream station is experiencing a statistically significant downward trend due to changing hydrological conditions. In contrast, the downstream discharge shows no clear long-term trend. Although both stations show negative slopes based on Sen's Slope Estimator test, only the upstream station's decline is significant, indicating that hydrological changes are initially detected upstream before they influence downstream conditions.

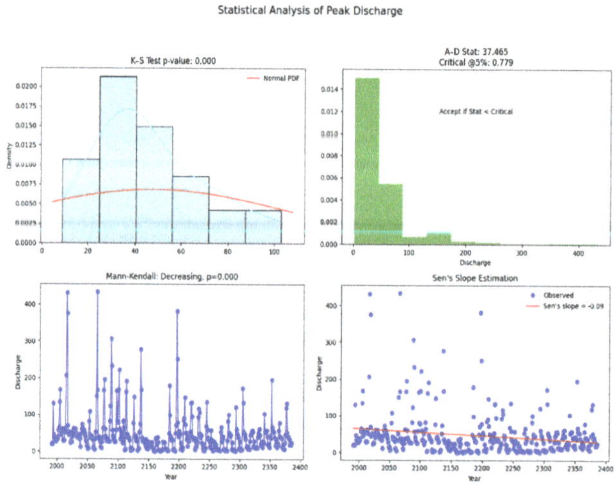

Figure 18. Statistical analysis of river peak discharge for upstream-33023 station.

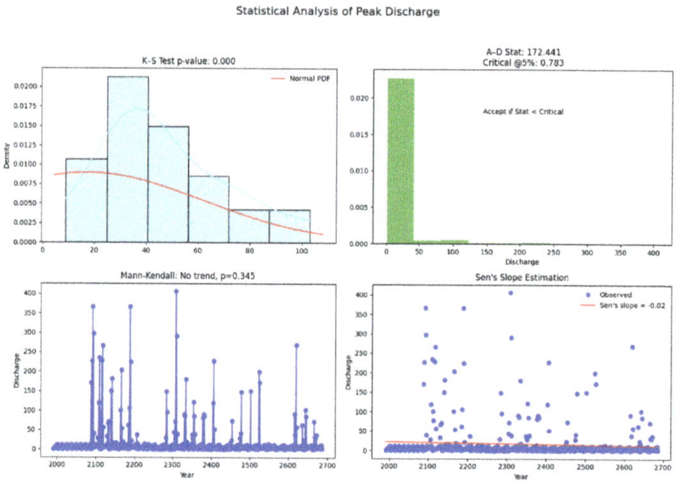

Figure 19. Statistical analysis of river peak discharge for downstream-33917 station.

The overall results indicate that the annual hydrographs for both stations reveal significant variability in peak discharge over the years, with notable reductions in downstream flow variability that are likely attributable to the regulating effects of the Bukan and Nowruzlu dams. The upstream-33023 station demonstrates greater sensitivity to seasonal precipitation and temperature changes compared to the downstream station, reflected in more pronounced fluctuations. The regression analysis of the monthly peak discharge indicated strong correlations with specific periods: September–October for the upstream site and December–January for the downstream site. These findings are consistent with the hydrograph behaviour, where the upstream discharge peaks earlier due to its proximity to the Bukan Dam, while the downstream discharge reflects cumulative contributions from the watershed, further moderated by the Nowruzlu Dam. Regarding precipitation, the upstream station exhibited peak precipitation during April–May during the 2000s and 2010s, whereas the downstream station showed a consistent shift towards May–June, with the 2000s presenting the strongest correlation ($R^2 = 1$). This temporal offset highlights the spatial and seasonal variations in rainfall distribution across the basin. The temperature regression analysis also revealed seasonal and decadal shifts: the upstream-33023 station demonstrated dominant temperature peaks during September–October (1990s and 2010s) and November–December (2000s), while the downstream-33917 station showed stronger correlations during the summer (July–August in the 1990s) and spring months (April–May in the 2000s and 2010s). These trends suggest that temperature plays a significant role in influencing snowmelt dynamics and evaporation rates, thereby affecting the timing and magnitude of the river discharge. Furthermore, the statistical distribution analysis confirmed that the Weibull distribution best fits the upstream-33023 peak discharge data, while the Lognormal distribution provides the best fit for the downstream-33917 data. This study contains regression analyses (via R^2 values) and goodness of fit statistical tests (via p-values) to identify the most critical months for flood forecasting and to recommend suitable statistical distributions for future modelling, respectively.

4. Conclusions

The historical data from the Zarrinehrud River for three decades from 1986 to 2018, including monthly river peak discharge (flow data), precipitation and temperature (climatic data), were used to investigate the river's behaviour. Due to the rising pace of climate change, human activities and the changing weather paradigm, flood management must be updated to revise the current management method. The upstream-33023 station exhibits earlier precipitation and discharge peaks compared to the downstream station, highlighting the role of topography and proximity to the Bukan Dam. The stronger correlations between precipitation and discharge at the downstream-33917 station suggest cumulative contributions from tributaries and seasonal rivers, compared to the more localised influences at the upstream-33023 station. The highest monthly temperatures at the upstream-33023 station occur during June–July, aligning with the peak of summer. Additionally, in the downstream region, temperature increases can affect the timing and rate of snowmelt from tributaries, contributing to discharge peaks in December–January. As both hydrometric stations are located at a considerable distance from the dam, the direct operational impacts have been minimised in the discharge measurements. Although detailed operational records were unavailable, the statistical analysis of the discharge behaviour remains a valid approach for understanding the managed hydrologic regime for the river system. It is recognised that dam operations, particularly those related to the Bukan dam, significantly influence the natural flow regime of the Zarrinehrud River, impacting the direct correlations between precipitation, temperatures and discharge records. Nevertheless, the objective of this study is not to isolate natural hydro-climatic effects from anthropogenic controls but rather to statistically analyse the historical river discharge, including the effects of dam operation. Applying statistical goodness of fit tests (K–S, A–D) and trend analysis methods (Mann–Kendall and Sen's Slope) provides valuable insights into the evolving discharge dynamics, capturing both the natural variability and managed flow conditions. Thus, these findings are highly relevant for decision-making under the integrated influence of climate and anthropogenic factors. Consequently, the analysis framework presented in this paper remains applicable for flood forecasting and adaptive management, even under managed river conditions. For further studies, it is recommended to integrate this analysis approach with detailed operational data, where available, to further enhance projections and inform proactive flood mitigation strategies.

Author Contributions: Conceptualization, F.M. and J.H.P.; methodology, F.M. and J.H.P.; software, F.M.; validation, F.M. and J.H.P.; formal analysis, F.M., J.H.P. and O.M.; investigation, F.M., J.H.P., Y.G. and P.R.H.; resources, J.H.P.; data curation, F.M., J.H.P., Y.G., P.R.H., M.M., E.A.-Q. and M.A.M.R.; writing—original draft preparation, F.M.; writing—review and editing, J.H.P., Y.G., P.R.H., O.M.,

M.M., E.A.-Q. and M.A.M.R.; supervision, J.H.P. and Y.G.; project administration, J.H.P.; funding acquisition, J.H.P. All authors have read and agreed to the published version of the manuscript.

Funding: This research received no external funding.

Institutional Review Board Statement: Not applicable.

Informed Consent Statement: Not applicable.

Data Availability Statement: The raw data will be made available on reasonable request to the corresponding author. This is because this is an ongoing study, and the authors need to keep a record to whom the data has been shared with.

Conflicts of Interest: The authors declare no conflict of interest.

References

1. Di Baldassarre, G.; Sivapalan, M.; Rusca, M.; Cudennec, C.; Garcia, M.; Kreibich, H.; Konar, M.; Mondino, E.; Mård, J.; Pande, S.; et al. Stoichiology: Scientific challenges in addressing the sustainable development goals. *Water Resour. Res.* **2019**, *55*, 6327–6355. [CrossRef]
2. Kazemi Garajeh, M.; Haji, F.; Tohidfar, M.; Sadeqi, A.; Ahmadi, R.; Kariminejad, N. Spatiotemporal monitoring of climate change impacts on water resources using an integrated approach of remote sensing and Google Earth Engine. *Sci. Rep.* **2024**, *14*, 5469. [CrossRef]
3. Kordi, F.; Hamzeh, S.; Atarchi, S.; Alavipanah, S.K. Agricultural product classification for optimal water resource management using the data time series of Landsat8. *Iran. Ecohydrol.* **2018**, *5*, 1267–1283.
4. Kazemi Garajeh, M.; Wenget, O.; Haghi, V.; Li, Z.; Garajeh, A.; Salmani, B. Learning-based methods for detection and monitoring of shallow food-affected areas: Impact of shallow food spreading on vegetation density. *Can. J. Remote Sens.* **2022**, *48*, 481–503. [CrossRef]
5. Plate, E.J. HESS Opinions "Classification of hydrological models for flood management". *J. Hydrol. Earth Syst. Sci.* **2009**, *13*, 1939–1951. [CrossRef]
6. Asadi, S.; Mousavi, S.J.; López-Ballesteros, A.; Senent-Aparicio, J. Analyzing hydrological alteration and environmental flows in a highly anthropized agricultural river basin system using SWAT+, WEAP and IAHRIS. *J. Hydrol. Reg. Stud.* **2024**, *52*, 101738. [CrossRef]
7. AghaKouchak, A.; Norouzi, H.; Madani, K.; Mirchi, A.; Azarderakhsh, M.; Nazemi, A.; Nasrollahi, N.; Farahmand, A.; Mehran, A.; Hasanzadeh, E. Aral Sea syndrome desiccates Lake Urmia: Call for action. *J. Gt. Lakes Res.* **2015**, *41*, 307–311. [CrossRef]
8. Mohammadi, F.; Mhammadi, M.; Behmanesh, J.; Jalili, S. Survey of Groundwater Level Fluctuations by Soft Computing Techniques (Case Study: Urmia and Miandoab Plains). *J. Irrig. Water Eng.* **2022**, *12*, 200–219.
9. Chen, H.; Guo, J.; Xiong, W.; Guo, S.; Xu, C.Y. Downscaling GCMs using the smooth support vector machine method to predict daily precipitation in the Hanjiang Basin. *Adv. Atmos. Sci.* **2010**, *27*, 274–284. [CrossRef]
10. Pu, J.H.; Pandey, M.; Hanmaiahgari, P.R. Analytical modelling of sidewall turbulence effects on streamwise velocity profile using 2D approach: A comparison of rectangular and trapezoidal open channel flows. *J. Hydro-Environ. Res.* **2020**, *32*, 17–25. [CrossRef]
11. Chen, J.; Li, Q.; Wang, H.; Deng, M. A Machine Learning Ensemble Approach Based on Random Forest and Radial Basis Function Neural Network for Risk Evaluation of Regional Flood Disaster: A Case Study of the Yangtze River Delta, China. *Int. J. Environ. Res. Public Health* **2020**, *17*, 49. [CrossRef] [PubMed]
12. Kumar, S.; Singh, H.P.; Balaji, S.; Hanmaiahgari, P.R.; Pu, J.H. Inclusive Hyper- to Dilute-Concentrated Suspended Sediment Transport Study Using Modified Rouse Model: Parametrized Power-Linear Coupled Approach Using Machine Learning. *Fluids* **2022**, *7*, 261. [CrossRef]
13. Toth, E.; Brath, A.; Montanari, A. Comparison of short-term rainfall prediction models for real-time flood forecasting. *J. Hydrol.* **2000**, *239*, 132–147. [CrossRef]
14. Te Linde, A.H.; Aerts, J.C.J.H.; Kwadijk, J.C.J. Effectiveness of flood management measures on peak discharges in the Rhine basin under climate change. *J. Flood Risk Manag.* **2010**, *3*, 248–269. [CrossRef]
15. Supriya, P.; Krishnaveni, M.; Subbulakshmi, M. Regression Analysis of Annual Maximum Daily Rainfall and Stream Flow for Flood Forecasting in Vellar River Basin. *Aquat. Procedia* **2015**, *4*, 957–963. [CrossRef]
16. Shirmohammadi, B.; Malekian, A.; Salajegheh, A.; Taheri, B.; Azarnivand, H.; Malek, Z.; Verburg, P.H. Scenario analysis for integrated water resources management under future land use change in the Urmia Lake region, Iran. *Land Use Policy* **2020**, *90*, 104299. [CrossRef]
17. Behboudian, M.; Kerachian, R. Evaluating the resilience of water resources management scenarios using the evidential reasoning approach: The Zarrinehrud river basin experience. *J. Environ. Manag.* **2021**, *284*, 112025. [CrossRef]

18. Kazemi Garajeh, M.; Feizizadeh, B. A comparative approach of data-driven split-window algorithms and MODIS products for land surface temperature retrieval. *Appl. Geomat.* **2021**, *13*, 715–733. [CrossRef]
19. Kumar, S.; Hanmaiahgari, P.R.; Chowdary, V.M.; Pu, J.H. Geomorphological characteristics of Gandak River between Sahibganj and confluence with Ganga River during 1989–2022. *Environ. Earth Sci.* **2024**, *83*, 501. [CrossRef]
20. Abbasian, M.S.; Najafi, M.R.; Abrishamchi, A. Increasing risk of meteorological drought in the Lake Urmia basin under climate change: Introducing the precipitation–temperature deciles index. *J. Hydrol.* **2021**, *592*, 125586. [CrossRef]
21. Khazaei, B.; Khatami, S.; Alemohammad, S.H.; Rashidi, L.; Wu, C.; Madani, K.; Kalantari, Z.; Destouni, G.; Aghakouchak, A. Climatic or regionally induced by humans? Tracing hydro-climatic and land-use changes to better understand the Lake Urmia tragedy. *J. Hydrol.* **2019**, *569*, 203–217. [CrossRef]
22. Heydari Tasheh Kabood, S.; Hosseini, S.A.; Heydari Tasheh Kabood, A. Investigating the effects of climate change on stream flows of Urmia Lake basin in Iran. *Model. Earth Syst. Environ.* **2020**, *6*, 329–339. [CrossRef]
23. Emami, F.; Koch, M. Evaluation of statistical-downscaling/bias-correction methods to predict hydrologic responses to climate change in the Zarrine River Basin, Iran. *Climate* **2018**, *6*, 30. [CrossRef]
24. Safari, N.; Zarghami, M.; Szidarovszky, F. Nash bargaining and leader–follower models in water allocation: Application to the Zarrinehrud River basin, Iran. *Appl. Math. Model.* **2014**, *38*, 1959–1968. [CrossRef]
25. SNCE. Studies of resources and expenditures of Bukan dam and Zarrinerud irrigation and drainage network. In *Sadrab Niroo Consulting Engineers*; 2012; 140p. (In Persian)
26. Katz, R.W.; Parlange, M.B.; Naveau, P. Statistics of extremes in hydrology. *Adv. Water Resour.* **2002**, *25*, 1287–1304. [CrossRef]

MDPI AG
Grosspeteranlage 5
4052 Basel
Switzerland
Tel.: +41 61 683 77 34

Atmosphere Editorial Office
E-mail: atmosphere@mdpi.com
www.mdpi.com/journal/atmosphere

www.ingramcontent.com/pod-product-compliance
Lightning Source LLC
Chambersburg PA
CBHW062041220926
51850CB00010B/2623
9783725873265